Building Rules

Building Rules

How Local Controls Shape Community Environments and Economies

Kee Warner
and
Harvey Molotch

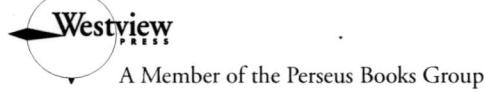

A Member of the Perseus Books Group

Some of the materials included in this book were published in earlier forms by the *California Policy Seminar* (Warner and Molotch, 1992) and the *Urban Affairs Review* (Warner and Molotch, 1995). Photographs are by Kee Warner.

Published in 2000 in the United States of America by Westview Press, 5500 Central Avenue, Boulder, Colorado 80301-2877, and in the United Kingdom by Westview Press, 12 Hid's Copse Road, Cumnor Hill, Oxford OX2 9JJ

Find us on the World Wide Web at www.westviewpress.com
A CIP catalog record for this book is available from the Library of Congress.

ISBN 0-8133-3923-5

The paper used in this publication meets the requirements of the American National Standard for Permanence of Paper for Printed Library Materials Z39.48-1984.

10   9   8   7   6   5   4   3

For
Katy and Eva Rose
Shana and Noah

Contents

Tables and Illustrations

Acknowledgments

We are grateful for the support of the California Policy Seminar, which funded our field research and published our initial report. We also appreciate the ongoing support of colleagues and staff of the departments of sociology and the libraries of the University of Colorado at Colorado Springs and the University of California, Santa Barbara. We had excellent help from a number of research assistants: Krista Paulsen, A. R. Lategola, Nancy Desser, Cathy Oakes, Mary Texiera, Caroline White, Rebecca Wypseck, and Ruth Jackson. Many colleagues contributed comments, critiques, and suggestions, including Richard Appelbaum, Hal Aronson, William Bielby, Bob Buster, David Diaz, T. Ryken Grattet, Bob Hughes, Rubén Martinez, Joseph Martorana, Max Neiman, Sergio Rey Jr., Robert Sollen, and Mark Shibley. We benefited as well from the comments of reviewers from the *California Policy Review,* the *Urban Affairs Review,* and Westview Press. We appreciate the professionalism of the Westview staff and their enthusiasm for this book.

Thank you to our local informants in the Santa Barbara, Santa Monica, and Riverside area who provided extraordinary commitment to this endeavor. We hope this book will be of value to them and to the many people working to build quality in communities and to protect natural environments. Deepest thanks go to our families for supporting this project from beginning to end.

Kee Warner
Harvey Molotch

1

The Relevance of Regulation

In the United States and much of the industrial world, business leaders, politicians, and large segments of the public question efforts to regulate economic activities for the protection of natural environments. Even though some environmentalists think the protections that have been enacted fall woefully short of creating sustainable human settlements, regulation critics attack the programs and policies that have been implemented as unjustified interference in the ways people produce and consume. Land use regulation—in particular the rules governing urban growth and development—is one particularly tangible terrain on which this battle is being fought. This book reports on just how this highly controversial form of intervention in the economy has operated in terms of either enhancing local communities and their environments or undermining their economies. Writing at a time when regulation is seen as the enemy of economies, we specify within the realm of land use and in the delimited but important region of the world we study—Southern California—just how regulation and markets have operated together. We illustrate what happens when the regulation of city building increases and the effects of this regulation on an economic sphere with major stakes for wealth creation and potential for far-reaching social and environmental consequences.

The creation of wealth in regulated contexts is, of course, not new—either in land use or any other realm. Markets presuppose the basic social order that guarantees property through contract; in the case of real estate, commercial value requires not only deeds backed by government force but also the infrastructure of roads and other forms of access that create economic utility. Going one step farther down the slippery but inevitable road of regulation, deeds and access mean nothing if their value can be destroyed by the acts of others, as when an upstream user diverts a common waterway or a neighbor performs activities so noxious that they seriously compromise the ability to use and enjoy one's own property.

In the economists' terms, actors in the marketplace can generate externalities, neighborhood effects, or spillovers.[1] To preserve the very order of the market, these offenses must be publicly controlled, either by preventing them in the first place or by otherwise providing relief for the victims—this is the "central problem of land use law" (Platt, 1996: 40). The U.S. Supreme Court sustained zoning laws originally in order to prevent a slaughterhouse from damaging the neighbors' land values. From the reasoning of that case, the way was cleared for a wide range of interventions whenever private land use had negative impacts on neighboring uses and, by further extension, on the community good.

Beyond the immediate stench of the slaughterhouse, we know that virtually all uses of land generate spillover effects, good and bad, near and far, complexly ecological or merely annoying to other human beings. These effects intersect in multiple and synergistic ways, over time and over distances great and small. If a homeowner cuts down a tree, the immediate neighbors have a poorer (or better) view; a landowner who clear-cuts can stimulate soil erosion that buries the neighbor's house under an eroded hillside. The loss of a unique local habitat can set in motion far-reaching ecological consequences. For example, Ellwood Grove, north of Santa Barbara, is one of a few overwintering sites for monarch butterflies that migrate thousands of miles throughout the western United States—a rallying point for environmentalists concerned with nearby development plans. No actor or even set of actors produces smog; it results when diverse people take actions that lead to the emission of specific chemicals that come together with sunshine and other natural elements to produce a qualitatively different phenomenon—a negative synergy occurs. These are the dramatic realities of Ecology 101. But virtually any human action on the land, mundane or spectacular, yields consequences that only our ignorance keeps from view.

Even with the field still in its infancy, ecological analysts are constantly uncovering the complex range of potential external effects—aesthetic, cultural, economic, and ecological—of human actions, and these ecological discoveries lead environmentalists to advocate new restraints on development. The "perpetual crisis," as Wendell Berry (1995: 65) puts it, caused by "our presence in this varied and fertile world ... forces upon us constantly a virtual curriculum of urgent questions." The ongoing and indeterminate character of understanding environmental impacts frustrates the linear-minded, who think that policy should be "done," that constraints on the economy should be minimal and settled, or that environmental protection should be, once and for all, accomplished. By requiring a specific environmental impact review (EIR) before individual projects involving federal lands or federal funding could be built, the 1970 National Environmental Policy Act (NEPA) institutionalized this indeterminacy, at least for certain developments. EIRs have become a tool for the continuous and contentious development of additional environmental knowledge, whereby the unfolding of answers comes from experts in various realms, but also from ordinary people directly understanding their environments by adopting scientific methods—"popular epidemiology," as it has been called in the case of residents' diagnoses of local environmental health hazards (Szasz, 1994).

Growth control has become a mechanism, perhaps a crude one, to deal with the perceived negative spillovers of urban growth. In its most simple and, as we will see, rare configuration growth control attempts simply to freeze the action. Developers liken this to pulling up the drawbridge. Stopping development, even cold and altogether, becomes one understandable alternative, especially when people do not understand the ecological consequences of development or when they have little faith that governments will properly mitigate the effects that *are* known.[2] The more common variations of growth control we describe in coming chapters embody a skepticism toward the development process and a break with past naïveté about spillover potentials.

Compared with the public sentiment about other topics, the concern for the environmental effects of development is not shallow. Surveys of American public opinion show continued widespread concern about pollution problems and the belief that government should play an active role (*Times Mirror*, 1990, 1994). According to the 1998 National Environmental Education and Training Foundation/Roper Survey, 62 percent of Americans believe that environmental protection and economic development can go hand in hand. If forced to choose one or the other, 71 percent would choose the environment (Coyle, 1998: 2). Majorities of people are willing to see more federal funding for control mechanisms (*Times Mirror*, 1990) and are willing to pay more for goods to better serve environmental goals (*Times Mirror*, 1994). Almost unique among the movement-based issues of the last decades, support for the environment cuts across boundaries of race, class, and gender and persists, even as support for other causes of both left and right (e.g., civil rights, balanced budgets) come and go (Van Liere and Dunlap, 1980). In a rare display of personal commitment, people also have changed some of their own habits, sorting garbage into separate containers, sometimes taking it on their own to recycling centers, and recycling enough waste to substantially reduce pressures on landfills (now approaching a one-third reduction in some communities). In no other realm of public life have individuals changed their behavior like this—usually for no reason beyond wanting to do the right thing. One of the first realms where conservative Republicans were forced to backtrack in their mid-1990s "Contract with America" was in their efforts to roll back environmental protection.

Growth rightfully takes the blame for mundane local difficulties, like effects on air quality and traffic congestion or just changing the "feel" or "character" of communities. People also perceive other problems, like crime, to be caused by growth; even though research does not support such reasoning, the belief helps build the movements against development. On more subtle ground, publics grow in their understanding that what were once taken as "acts of God" involve acts of development. Forest fires destroy homes built in the paths of what we now know to be events of natural regeneration; "out of control" water sweeps away whole towns built in flood plains by professional developers; in the arid West, overbuilding drains local water supplies creating "droughts" within the normal precipitation cycle, leading to the imposition of water conservation measures and to the high costs of securing alternate sources of water. Increasingly, these difficulties raise the possibility that to make development more harmonious with nature there should be changes in the way cities grow.

For all sorts of reasons, then, some profound and some trivial, some quite correct and others off the mark, some urged by national experts and some derived from local experience, environmental awareness has filtered into public consciousness. Although suspicions abound that intervention hurts society and economies, there is parallel suspicion that life, community, and earth can not be left merely to markets. We suspect that these two perspectives, however paradoxical, coexist not only in the same cities and communities but within many individuals. Clarifying the effects of regulation may help assuage dilemmas at both the levels of policy and individual consciousness.

Environmentalism as Local Urge

The measures we deal with are local—those who focus on global crises might say "merely local." But local policies have impacts on larger realms. Macro-economies, of course, result from concrete economic activities that must occur somewhere. The cumulative effects of local regulation add up to impacts for the economy overall, a point obviously not lost on the critics of regulation. But the same is true for cumulative ecological effects. The way cities grow determines, for example, the need for cars, how much they are driven, and the specific meteorological environments into which their exhaust will flow—including the amount of "greenhouse gasses" produced and their effects on global warming. The form settlements take shapes the use of all sorts of environmental resources, not only fossil fuels but land, habitats, water, air, and, on the other end, the production and disposal of the array of human "wastes." The particulars of settlement shape consumption of all sorts, including the very need for such durable goods as cars, lawnmowers, and air conditioners, and the capacity of land-fills to absorb the aftermath of consumption. The size, configuration, and political organization of cities influence the capacity for regionalism; the ability of particular regional settlements to be more self-sufficient diminishes broader environmental impacts. Local development—its overall scale, the materials used, the amount of energy required, the waste produced—reverberates through and around the earth.

The local scene is also an important setting for action and social change that can redirect these development processes; it makes sense that environmental politics begins in the backyard. Locality is the learning site and, by necessity, the place where new ways of life can be tried and witnessed. At the same time, the social, political, and economic organization of urban growth offers people a model for relations with nature more generally. True, distant catastrophic environmental threats to human life—like the nuclear crisis at Three Mile Island and the poisoning at Bhopal, India—also teach and inspire action. But such stories' impact comes not just from the fact that an agitated world press tells what happened, but from local populations' imaginings that such tragedies could happen "right here at home." News of wars among distant peoples, imagined as driven by primordial hatreds or schemes of tyrannical regimes, has only weak implications for those who think they live in stable democracies among more or less reasonable people. But people take distant human-made environmental crises to be applicable to their own communities. Chernobyl set back nuclear development across the democratic world, even where the technology was far more sophisticated and the monitoring systems more competent. Risk assessment studies show repeatedly that people fear most that which they can not control—for example, their airplane's safety compared to slipping in the shower (see, for example, Starr, 1969). They have special concern and anger over those risks they think others could and should have controlled—adding still another dimension to troubles perceived as coming from human-made development.

The local also matters, both as a force and site of environmental study, because it is the cauldron where diverse constituencies negotiate their competing needs and form coalitions. Race and class divisions create a challenge for any type of social movement, especially those without resources of great wealth and corporate power. A reform movement like environmentalism, which challenges major economic interests and those who depend on their campaign contributions, can afford to alienate few other constituencies. By embracing the concerns of low-income and minority populations, urban environmentalists can bolster "environmental justice" and address community environmental quality more effectively.

Inner city communities do bear the biggest urban environmental impacts; they are disproportionately bisected by freeways, built on land permeated by poisonous byproducts of industrialization, and more likely to be adjacent to locations chosen for garbage dumps and incinerators.[3] By defending inner cities as a vital part of the urban ecosystem, environmental organizations broaden their base. By preventing (or remediating) these hot spots of environmental degradation, they also control pollution more efficiently by addressing the problem close to its source. This is still another result of environmental knowledge: Source reduction protects those, human and otherwise, more distant. Naive and sentimental "Bambi projects" that secure the woods around a manor house do not become the primary focus of environmental reform. Responding to the needs of diverse local urban groups not only wards off a trivial environmentalism, but also helps build larger coalitions that go beyond environmental issues.

The most intense and repeated citizen resistance comes in reaction to specific developments—often dubbed LULUs (Locally Undesirable Land Uses). But the LULU category has become more inclusive over the past generation (Blumberg and Gottlieb, 1989; Hamilton, 1990; Piller, 1991). Newly awakened constituencies thwart plans not only for toxic dumps and prisons but also for what were once utterly uncontested types of development, like university campuses, private research facilities, luxury homes, and sports facilities. Whether as a cover for more selfish motives or because of an authentic vision (or both), citizens fight to keep wetlands intact or an endangered species' habitat undisturbed. A millionaire's house on wetlands becomes as much an offender as a public housing project in a single-family zone. The array of LULUs thus grows as communities define and defend a range of valued local qualities, including natural ecologies.

In our most immediate experience, the "environment" is in the breath of air we take, the view before our eyes, the noise that enters our ears, and the odors that reach our nostrils. Despite environmental organizations' ambitious involvement in national and international policy making, localities are where we sense the environment. When people are suspected of reacting to these sensate impulses, they can be accused by more practical business people (and even some environmentalists) of "emotionalism" or of not seeing "the big picture." An alternative is to view the quality of these experiences as the very essence of life, certainly as crucial in practice as any of the artificial products that come from defiling these experiences or imitating them through merchandise and commercial entertainments. These intrinsically local connections with "nature," whether based in reason, emotion, or both, offer fertile grounds for challenging conventional methods of building urban environments.

Wendell Berry—poet, farmer, and social commentator—warns that the celebrated bumper sticker "Think Globally, Act Locally" is too mechanical in its local-global split. It may distract environmentalists from the need to think *and* act locally in order to create necessary changes in the local and regional arenas where they can and should be effective (Creedon, 1993). The immensity of global issues can leave reformers defeated or can become a form of mere symbolic politics, waged through the convenience of bumper stickers, Sierra Club dues, and platitudes. Growth control groups, although typically short of Berry's depth of understanding, are—whatever else—focused on the local in both their thinking and their acting.

The results of such local thinking and doing have gone beyond any particular locality by dint of their volume. As early as 1975, over 300 U.S. jurisdictions had adopted some form of growth control (Logan and Zhou, 1989: 461; Cohn, 1979). By 1988, 150 localities in California alone had enacted control measures (Glickfeld and Levine, 1990: 5). By another count, environmental groups were sufficiently organized in about one fourth of all California cities to make significant impacts on growth policies (Clark and Goetz, 1994). The content of local policies ranged considerably in restrictiveness and scope, but, by one indicator, 10 percent of California jurisdictions had adopted explicit population growth caps by 1990 (Glickfeld and Levine, 1990: 23). Although exclusive suburbs have long engaged in some form of density and growth limits (indeed such was their raison d'être), this new wave of land-use control involves a far more diverse array of communities: old towns and new, central cities and fringe, white and minority, small and very large. Even where environmentalists did not take control—and this was certainly the case in the great majority of U.S. localities—tightened procedures did at least gain toeholds in many places, and with them came new types of orientations.

The fight for local controls has been a proving ground for people who then move on to state and national policy realms, both as activists in national groups and as members of state legislatures and federal agencies. The growth control movement thus not only has the potential for changing local policy but for altering the ideological stripe of state and national civic life. We return, then, to the issue of scale: By virtue of their content, repetition, and sponsors, the potential impacts of local controls become large—not always global but certainly well beyond mere parochial interest.

The Critique: Stranglers of the Economy

The perception that controls were indeed becoming commonplace, moving beyond the traditional suburbs of the exclusionary rich to a more

general urban order, helped bring on the mid-1990s reaction of deregulation and government downsizing. In the United States, a need to build
the economy dominated the news, particularly in prior high-growth
zones like California—also the leader of the country's environmentalist
reforms. Following an increasingly pervasive economistic logic, some deduced that when localities limit either the amount or the conditions of
development, some businesses may be killed off altogether—taking with
them jobs and needed housing. By the same reasoning, if environmental
rules raise production costs at factories, the goods will be priced higher
and markets will be lost to places abroad with more favorable business
climates. Jobs and corporate tax dollars will evaporate, leaving unemployment, welfare costs, and homelessness in their wake. So runs the
newly invigorated argument for freeing up the hidden hands to have their
way with the social and physical landscape.

Even in the former strongholds of environmentalism, the language of
"rebuilding the local economy" came to replace, or at least overlap, the
expressed concern for quality of life and environmental protection.
"Business" became the buzzword for political candidates of virtually any
stripe; in the 1994 California elections, nearly five times as many candidates listed their occupations with business designations (e.g., "business-
woman, businessman") with the next largest occupation being teacher or
school official.[4] Designations like "attorney" and "public official" (much
less "environmentalist" or "activist") fell from usage, as politicians learned
to play up even marginal business roles as the best identity for a ballot
victory.

Government leaders, beginning with President Bush but including California's Governor Pete Wilson and other leading Republicans, competed
in denouncing government's stranglehold on private enterprise. Even
Wilson's Democratic opponent for governor, Kathleen Brown, ran on the
platform that Wilson's rule was responsible for California having "the
country's worst economy"—a very dubious claim but evidently one that
her strategists thought would be effective (Stall, 1994: A3). Blue-ribbon
commissions cautioned that "overregulation" stymies urban economies
(Advisory Commission on Regulatory Barriers to Affordable Housing,
1991: 1). They denounced land-use and environmental controls, in particular, for their nefarious effects on local and national economies as well as
on the chance for the middle class to have jobs and the poor to survive.

On the consumption side, critics blamed regulation for raising housing
costs, especially in places like the coastal California cities that experienced radical price inflation beginning in the mid-1970s. By curtailing the
supply of land available for housing, the argument went, the price of the
housing that remains would go up. This price effect on workers' housing,
as well as the general drag on the economy, caused California's "Ueberroth

Commission"—the group appointed by the governor to bolster the state's flagging fortunes—to echo the same refrain against "excessive rules, regulation, and red tape." Peter Ueberroth, a travel agency business magnate, had gained celebrity through heading up the Los Angeles Olympics, the games that came in with a profit as well as the usual basket of medals for the American teams. In later appointing Peter Ueberroth as head of "Rebuild-L.A."—the blue-ribbon organization that was supposed to deal with the problems the city faced in the wake of the post-Rodney King insurrections—Democrat mayor Tom Bradley received the now usual prescription: a recommendation for "clearing away" unnecessary regulations to stimulate development in the city's most deprived zones. In this way the poor would get jobs and the city would have its peace.

Such reasoning has widespread appeal, even within groups like professional planners who have long complained of developers' influence. Planners can come to resent the comparatively privileged middle-class neighborhood residents who cry "Not in my backyard" (hence, "NIMBY") as they use larger planning and environmental talk to protect their allegedly narrow turf. As with the developers, these groups may also interfere with professional planning goals (such as increasing building densities around transportation nodes) and hence irk those who otherwise might be in their alliance. Citizen groups also can fall into class and racial exclusivity, sometimes leading them to oppose projects that provide jobs and housing for people of lower social standing.

In a move linked to the Republican's 1994 "Contract with America," some in Congress proposed government compensation to any private property owner whose land or building values were lowered by regulation (no mention was made of special payments *from* owners whose values were *raised* by public regulation—likely to be a frequent occurrence). Although unattained at this writing, such a measure—one the courts have come increasingly close to enacting through judicial rulings—would radically curtail even zoning controls, not to mention the kind of project impact remediations that have become increasingly commonplace.

As with the Ueberroth reports, the case for deregulating land development is usually made on the basis of the jobs it supposedly brings, the costs of housing it presumably lowers, and the strengthening of the tax base it putatively engenders. These are the up-front, manifest reasons. But behind such high-sounding concerns are the pecuniary interests of particular groups who gain returns from real property. Growth is their bread and butter; the more intense development can become, the greater the fortunes they can make. The stakes can be huge; whereas returns on some other types of business ventures may be on the order of 10 or 15 percent, the successful approval of an office development on, say, an agricultural parcel can generate a five, ten, or even 100-fold increase in market value just because a more intensive use is being allowed. If speculators buy property on the assumption that local officials will approve a higher use, a project denial can be devastating. No wonder they devote such time and resources toward affecting regulatory environments and the daily operations of local government.

When it comes to particular projects, people in the growth business find restrictions of any sort a pain in the neck, at best, and, at worst, a potential source of business crisis. Auxiliary professionals such as the lawyers, architects, accountants, civil engineers, and others who specialize in real estate development provide sympathetic and often detailed backing. Supported politicians express the public-spirited version of developers' concerns as do the local media (who rely on aggregate local growth for profits). To various degrees, other types of business owners come on board, some because—like the media—they perceive future profits as tied to aggregate local growth, and others through bonds of social and ideological solidarity.

This cumulative set of forces that powered local growth machines in most of the United States before the environmental era helps explain why there has been such inconsistent enforcement of zoning laws in U.S. cities over the years (except in the richest residential areas). Finding more lucrative uses of property was usually sufficient reason for changing the zoning (Babcock, 1969)—sometimes, of course, such changes required that developers make a well-placed campaign contribution or simple bribe (what the Italian developers call *bustarella*—literally, "small envelope"). These cumulative forces also explain how "farmer-developers" converted rural lands to suburban uses so easily, regardless of the ecological waste and fiscal inefficiency of the resulting sprawl (Rudel, 1989, Markusen, 1978). In a far-reaching version, the sway of growth promoters helps explain even the collapse of South American rain forests (along with the Florida everglades), as state action and the demands of local land speculators for highways and police protection push the development boundary outward into wilderness (Rudel and Richards, 1990; Rudel and Horowitz, 1993). Forest protection, as is often the case with conservation goals more generally, requires understanding the networks and coalitions that promote growth—the supply side, if you will—rather than simply the demand forces—whether these be peasants seeking subsistence plots, or buyers for new suburban housing.

Whether in tropical rain forests or the cities and suburbs of industrialized nations, investments in land and buildings give certain actors both the motivation and resources to dominate the production of places. At least in the U.S. context, the leadership of these actors rests on a bias built into civic culture, legal precedents, and economic institutions. As a result, cities have operated as "growth machines," with local governments serving as instruments to facilitate still more development.[5] Governmental capacity always operates as part of a particular "regime," which includes the surrounding economic and ideological context that sustains formal authority.[6] The growth machine system is a particular kind of local regime, one that robustly characterizes American cities throughout their histories.

One obvious caveat is that the pro-growth coalitions, as with their adversaries, are not themselves utterly coherent; and, although our point of departure has been to stress their inherent advantages, they do have various fault-lines that weaken their effectiveness and help provide slack within which environmentalists can operate. Sometimes, for example, land-use elites may quarrel over the costs and benefits of infrastructure (e.g., dams and highways)—just who will pay the dollar costs, who will gain the building contracts, or who will prosper by floating the finance bonds. The more usual conflicts, quite tellingly, center on the location rather than the amount of growth: Whose property will gain by being near the airport or convention center (Banfield, 1961)? Or which land region will gain water rights from some far-away source (Hundley, 1992)? Although land-use elites may share a desire to "grow" the community as much as possible, the distribution of growth—determined by infrastructure location—becomes the bone of contention.

In some cases, property interests may turn against a development project that undermines a favored growth scenario. For example, local elites of central and northern California cities opposed oil production offshore as threats to tourism and related development. Casino and resort owners in Lake Tahoe have allied with environmentalists to promote development that will not muddy Tahoe's crystal waters quite so quickly—provoking the rage of "dirtier" tourist concerns (Christensen, 1997). Land-use elites in Orlando, Florida, locked horns with Disney World for hogging too many of the spillover money benefits of the amusement center—that is, hotel, retail, and residential development (Foglesong, 1990). Local businesses have even urged a form of growth management as a means of limiting the scale of Disney operations so as to gain greater access to the park's side-bounties for local enterprises.

Although rare on the scene, some businesses may have opposed additional growth because they feared introduction of new competitors. At one time, elites involved in manufacture (as opposed to real estate) may have thwarted new factories that would compete for local labor (Smith and Keller, 1983). Although the historic evidence is scant that this practice was widespread or effective (see Scheiber, 1973), some contemporary real estate operators do, on occasion, try to block additional projects that would compete with them for tenants or buyers. In Ventura County, California, the owners of a regional shopping center tried to block construction of a competitive mall by bankrolling a referendum to eliminate the local government subsidies behind the project (as well as all other such subsidies). Although doomed to failure through stiff opposition from local government and business groups who feared for "the city's ability to compete with other cities," the effort does signal the kind of tension that can occur even within the world of local real estate (Weiss, 1996: A3). Another bone of contention has arisen when a city singles out certain developers or corporations for special abatements; those left out have litigated with the result that procedures are established to provide the abatements to all who meet standard threshold criteria (Wolkoff, 1983, as cited in Eisinger, 1988: 152). But the criteria inevitably favor some businesses (usually the largest corporations, which have in fact gained the most subsidies) at the expense of others.[7]

As usual, then, politics can make for strange bedfellows when the occasional dissension within the growth groups takes hold. But the historic record, we emphasize, shows this to be the rare event. Business operates with remarkable consensus on the growth issue, and environmental groups ordinarily operate in opposition to the goals of business. "Economy versus environment" has become the generic frame, with the two sides clear enough on most every battle and dispute.

Economy Versus the Environment:
Some Past Findings

We return, then, to a central focus of this inquiry: What is the nature of conflict between environment and economy? Rather than accepting any narrow usage of these terms, some thinkers, like Wendell Berry and Wes Jackson, lay out the utter confusion of ever regarding economies and environments as being in contradiction. The real economy, or what Karl Polanyi early in this century called the "empirical economy," can never be in conflict with quality of life; indeed if we put aside the models deduced from conventional microeconomics, quality of life, including environmental benefits for future generations, becomes the key indicator of economic well-being. On intellectual grounds, an increasing number of writers (for example, Beatley and Manning, 1997; Daly and Cobb, 1989; Hawken, 1993; Shiva, 1988) thus reject the dichotomization of environment and economy and suggest these realms must be reintegrated to build sustainable communities. Such organizations as Redefining Progress are attempting to replace traditional economic measures such as the Gross Domestic Product with more subtle and comprehensive indicators of economic health and community quality (Cobb, Halstead, and Rowe, 1995). Although these approaches are gaining momentum, they still have had limited impact on local discourse regarding development and growth.

Environmental activists often argue their case within conventional framings, but ironically this is precisely where the evidence of a conflict between the economy and the environment is far more ambiguous than the usual talk presumes. There are good reasons to doubt the intrinsic economic benefits of urban growth, even in the narrow sense of improving such standard economic indicators as unemployment rates, public budget viability, or levels of retail sales. Business groups argue that communities should accept a certain level of environmental damage as a necessary byproduct of urban growth; such damage is simply the price of maintaining a healthy economy. But there are reasons to doubt whether there truly is a payoff, even in economic terms.

Despite the drumbeat of enthusiasm for growth by politicians and consultants and an occasional study showing its benefits (see Muller, 1975), the great thrust of research questions whether local economic development strategies actually yield jobs or deliver any other net benefit: Is the "creation" of jobs offset by the influx of new job-seekers and/or the costs of the local incentives that have been paid out? According to one series of studies, localities with rapid development (measured in a variety of ways) experience no lower rates of unemployment than localities that grow more slowly (Appelbaum et al., 1976; see also Sternlieb and Hughes, 1983; Fasenfest, 1986). Some evidence suggests an arc-shaped curvilinear relationship between economic vitality and the rate of population change. Communities with the highest rates of population change, whether growing or shrinking, are at the tails of the curve and have the worst economies. The strongest economies exist in localities toward the top of the curve that are the most stable in terms of population size.

Studies of prominent cases tend to present a critical view of growth and its net benefits. There is the case of Atlantic City, New Jersey: a dream scenario of robust redevelopment under the aegis of the gambling industry, but with the jobs captured by migrants and suburbanites, not the struggling residents in whose name the casino projects had been launched (Sternlieb and Hughes, 1983; Williams, 1995).

In terms of the fiscal advantages of growth for local governments, the data are consistently negative (Appelbaum, 1978; Danielson and Doig, 1982; Eisinger, 1988:44): Growth tends to generate costs that exceed increases in local revenues, although the kind of growth matters (rich residents pay more in property taxes than the poor because the homes of the rich have higher market value; some industries generate higher costs than do others). In reviewing three major studies of the fiscal impact of sprawl versus managed growth, Burchell (1997: 171) reports the clear fiscal benefits in terms of road and utility infrastructure of managed growth versus unregulated sprawl, with more "modest' savings on school construction. Of course, local governments may profit in fiscal terms if they can attract high yielding enterprise with workers who live in a different jurisdiction (though this hardly increases the general community or environmental good). The careful analyses of specific projects that environmental impact review may entail often reveal that new developments do not pay their own way. The costs they generate in road maintenance, traffic management, and school services outweigh the taxes they generate. The new projects offer no real solution to local unemployment, either because the skills needed are not local, or because high profile economic development efforts motivate more people to move into the community than there are new openings in the labor market. Our point here is not to provide a single bottom-line from the research literature but to show again that scientific research does not demonstrate growth to be of net economic benefit. This murkiness of economic consequence contrasts with the more clear-cut findings on the environmental and quality of life costs of development.

At a more basic level, it is uncertain whether efforts to lure development pay off in the still more confined sense of attracting new projects at all. It may be that efforts to create a "good business climate" at the expense of environmental quality and social equity actually do not work: Places that attempt to create such a climate merely establish subsidies for phantom developments that never materialize, or underwrite ventures that would have happened on their own. Several careful studies examine the fate of those localities with the best "business climates"—low taxes, low spending, weaker environmental controls, fast growth rates, pro-business "receptivity"—based primarily on standard ranking sources from within the business world such as *Inc.* magazine, Fantus, and Grant. In comparing states' economic performance, Freudenburg (1993) determined that, if anything, a good business climate predicted bad economic outcomes in subsequent years. A comparable study by Skoro (1988: 151) concluded that business climate indicators "are useless as predictors ... [and] worse than useful as guides to state and local government action." After conducting still another comparison, Meyer (1993, forthcoming), concludes that states with more intense environmental standards, by most indicators, performed better economically in successive periods than states with weaker standards. At minimum, says Meyer (1993: 10), "we can conclude that shifts in environmental policy, whether intended to extend environmental control or reduce it, have no discernible effect on state economic performance."

Using very different methodologies, economists have tried to discern impacts on manufacturing and trade caused by environmental restrictions in various countries—that is, they have tried to track whether or not industries, especially polluting ones, escape to "pollution havens." The repeated finding is that costs of complying with environmental controls are simply not consequential enough to have any detectable effect on corporations' behavior (Leonard, 1988; Tobey, 1989, 1990). Such costs as labor and raw materials overshadow environmental considerations, which means, according to still additional studies (McConnell and Schwab, 1990; Bartik, 1988) that firms' location decisions likely have little to do with levels of environmental controls. Economists Cropper and Oates (1992: 699) summarize the policy implications: "There is little force to the argument that we need to relax environmental policies to preserve international competitiveness."

Some studies use corporate officials' own survey responses to specify whether or not pollution abatements, regulatory relief, and other forms of assistance make a difference in their location decisions. Some reports amount to little more than journalistic collections of "horror stories" (Frieden, 1980; Plotkin, 1987), but some have been conceived more carefully. Given corporations' ongoing ideological commitment to minimal regulation and their fondness for local "incentives" that subsidize development, this method intrinsically biases results in favor of low regulation. Nevertheless, even these findings are mixed. Eisinger (1988: 220) summarizes the findings regarding tax and financial incentives—often thought to be the most powerful form of inducement—by saying there is "little warrant in these various studies to argue that [these incentives] play a role of much, if any, significance in plant-location decisions," although "some evidence" does suggest that a favorable overall tax structure on business may have some effect.

Other studies using towns as the units of analysis (instead of states or nations) indicate that local governments' efforts to attract projects are usually ineffective (Humphrey and Krannich, 1980; Summers, 1976). Either the plants the towns get would have come anyway, or local governments spent money and effort to woo projects they never really had a chance of landing. Even when towns do win projects, the victory can be costly. Case studies of particular projects—like the city of Detroit's subsidies of General Motors to create the "Poletown" project (Fasenfest, 1986), and Flint, Michigan's, public-private investments in convention and amusement facilities (its "Auto World" project)—provide examples that show unhappy financial results for some cities (for a series of cases, see Squires, 1989).

Evidence suggests that enterprise zones—which free up targeted urban areas from environmental and other restrictions through tax abatements, pollution control relief, public subsidies of new hires, and loosening up zoning and building restrictions—do not lure new business or generate new jobs. The trivial numbers of jobs such zones do create are seldom taken up by local residents (Battle and Underhill, 1986; Eisinger, 1988: 194–199; Dowall, Beyeler, and Wong, 1994). To put it in its most foreboding form, localities merely compete with one another in a "race to the bottom;"[8] overall social and environmental welfare is sacrificed without even minimal local economic gains.

The fact that local efforts to simulate development generate so little positive economic impact suggests that local policies, whether pro- or anti-development, may have little impact on what goes on—in either direction. In one landmark study, Logan and Zhou (1989) found that growth rates of cities with growth controls were no lower than those without them. Molotch and Louch (1994) compared Southern California counties and metropolitan areas during the recession of the early 1990s and found that localities with strict regulations did no worse (perhaps even better) than localities with more open policies; at least during recession, the regulatory environment had no apparent impact. In his search for impacts of growth controls on housing costs, Landis (1992) found that a group of California growth control cities had no higher rates of housing price inflation than did otherwise "matched" cities that lacked such controls—implying perhaps that the controls had not, in fact, lessened supplies. There are, in these cases as well as in certain other studies with different results (e.g., Feiock, 1994), particular technical issues that may help explain the discrepant findings (see Appendix A). Our point here is that the research, taken as a whole, provides good reason to question some of the common assumptions about the economic impacts of either promoting or limiting local development. There may be impacts, but to find them will require more subtle analysis of how regulation works.

Local Regulation and Urban Theory

Understanding the forces that determine local policies and their outcomes has been a long-standing focus of scholarly research and debate. The dominant 1950s and 1960s debate pitted "elite" versus "pluralist" interpretations of power. The former construed cities as dominated by a single, coherent corporate-based elite, whereas the latter portrayed urban power as shared within a coalition of interest groups, sufficiently diverse in membership to reflect a form of democratically dispersed participation.[9] The contest ended not so much with a victory of one side over the other as with the growing realization that the study results were primarily an artifact of the method and the discipline of the researcher. Sociologists typically confirmed domination by something like a power elite, whereas political scientists learned that power was plural (Walton, 1966). In reviewing this result, Walton concluded (1976) that the way to escape the stalemate was to look beyond the *structure* of community power to its tangible *consequences*, like social inequality and who gets what out of particular governance processes.

A generation of new urban scholars moved beyond the simple dichotomization of plural politics versus a single economic elite (for reviews, see Zukin, 1980; Walton, 1992; Gottdiener and Feagin, 1988). In the "City as a Growth Machine," Molotch (1976: 310) posited that growth provides, in substantive terms, the central unifying political agenda for local elites in the United States, however divided they might be on other issues. This fact skews public benefits in the direction of local elites and affects the life chances of others, usually hindering them. Under elite dominance, growth promotion becomes the critical function of local government, and because it so dominates agendas, it operates as a key "constraint on initiatives for social and economic reform" (310). Logan and Molotch (1987) elaborated by describing how those principally concerned with increasing the value of their investments in real estate tend to overwhelm those oriented toward use values of the city, values stemming, for example, from attachments to community or from just striving to find a place to live. A stream of research and commentary explored these issues in larger detail (for overviews see Logan, Whaley, and Crowder, 1997; Jonas and Wilson, forthcoming).

Although also highly suspicious of the pluralist conception of the city, other refinements of the elitist paradigm have been less centered on growth elites compared to other groups in the polity. Urban "regime theory" takes local government institutions as sufficiently important and complex to force elite accommodation with at least some other types of actors. Hence although business interests, both corporate- and real estate-based, are central—maybe even the most important—they must have coalition support from at least some other groups to move their projects forward. Thus, an urban regime must contain a coalition of groups controlling different types of resources in order to advance a particular agenda (see Elkin, 1987). Trade-offs are made between those who have access to different types of organizations, to different ways of adding to the tax base, and to ways of gaining votes of different constituencies, and so forth. Stone (1989), for example, shows how Atlanta's downtown business interests facilitated the incorporation of African-American leaders into the governing coalition, albeit within a development agenda that favored downtown interests. Although urban regime theorists stress the need for coalitions, they stop short of depicting them as pluralistic gatherings of "equals at the table." Indeed Elkin (1987: 36–37) suggests that efforts to promote economic growth are one of the three defining axes of local politics. He goes on to say that "promoting economic growth in the city has come to mean viewing the city as a pattern of land use"; this way of putting it comes very close—despite the emphasis on coalitions—

to matching the assumption of growth machine theory (see also, Logan, Whalen, and Crowder, 1997: 607).

Still another variation of the local power debate comes from those who emphasize the larger national and global contexts within which any local power can be exercised (see, for example, Lauria, 1997; Feagin, 1998; Sites, 1997). Under this view, localities must contend with the broader transformations that switch economic activities on or off (e.g., manufacturing versus service economies) and, in so doing, create the conditions for local governance. This means some areas undergo deprivations that may feed disruptions if not handled or "regulated" in some way. This may mean police crackdowns or some combination of social programming to stave off electoral uprisings or protests in the streets. With varying degrees of success and imagination, repression or entrepreneurial spirit, localities strive to find a viable niche. At the same time, these outward-looking strategies shape the conditions for local regime-building and substantive policy.

Given the steady expansion of globalizing trends, some argue that local leaders have less and less room to maneuver (Gottdiener, 1987; Zukin, 1991; but see Cox, 1992). On the other hand, highly localized surges of religious nationalism and/or place-based identity fuel resistance to globalization in diverse spots around the world. These surges suggest that locality is a persistently autonomous realm of action. Local insurgencies are causing larger units to collapse, national boundaries to be redrawn, old patterns of governance and trade to be modified. As Giddens (1985), among others, points out, the traditional nation-state has lost out, both as a center of power and as a sensible unit of analysis. Neighborhoods, towns, and regions are the more tangible loci of human action that are directly woven into globalized institutions. Globalization may even heighten the importance of locality as this becomes the place where all the global interconnections are given form and define sites of investment, spaces for political action, and stages for cultural transformation. Whether in the former Yugoslavia, the villages of Chiapas, the disputed boundaries between Israelis and Palestinians, or in the deindustrialized urban economies of the United States, the combatants (on all sides) wear the marks of global culture and trade: They turn their faces to the global television networks wearing the latest youth fashion, "tribal" tattoos, and other appurtenances of global consumption and taste systems. Lefebvre's formulation seems radically apt: "No space disappears in the course of growth and development: the worldwide does not abolish the local" (Lefebvre, 1991: 330). In the U.S. context, although localism may not take the form of sustained violent insurgency, the country's particularly decentralized system of land-use control may offer special opportunities for local initiative.

But even if local constituencies demand control of growth and if governments have the technical means to accomplish such purposes, does this override the traditional hegemony of growth and business elites? Within localities and across a wider terrain, economic, state, and cultural institutions offer power advantages to those best situated, and so power grows on itself, influencing the urban ecology. But advantages do not always translate into outcomes, and certainly not always smoothly. Hegemony, as Stuart Hall has remarked, takes work, and the process can be complex. Finding that work in real contexts that matter, like land development, becomes a means for understanding power and contingency more generally.

We see the locality as the small end of a local/global linkage. Here new powers are claimed and even reinvented, as evidenced by locally based social movements around the world organizing for the environment, democracy, and cultural autonomy (Esteva and Praleash, 1998). In explaining the grandest transformation of our time, Feshbach and Friendly (1992) say local environmental movements were a key source of the undoing of the Soviet Union. Even in that severely constrained context, localities apparently contained the potentials to reorder political economies—potentials stirred up by people's desires to better control their physical environment and routine way of life.

We recognize the relevance of all these theoretical concerns, but joining an effort to find the fine lines that discriminate among them would not advance our present goals. We can acknowledge that, in addition to growth elites, others play important local roles and still maintain our focus on growth processes as crucial. We can see that global transformations influence local possibilities, even as we examine the consequences local policies actually have. We can ask whether policies intended to protect local qualities and citizen interests are economically disastrous within a global context. If they are, it will be more difficult to withstand new external pressures, otherwise the potential for local action and policy innovation may be much broader.

Growth machine theory, which we continue to find useful, is an analytic tool. It is not a deterministic doctrine in which growth is a preordained biotic or political imperative driven by the smooth operation of land markets or the unassailable power of an elite group. It is a "more or less" and somewhat contingent set of propositions. The point of this study is not to prove or disprove its central hypotheses, nor to re-specify who makes up growth elites or how those elites come to power and relate to other groups. That is work for other projects. Here we are focused on what follows the winning of elections, the adoption of voter initiatives, and the ideological combat associated with such events. We ask how political shifts on the growth control front play out in the project-by-project and rule-by-rule regulation of development, and with what consequences. This focus provides a glimpse of pivotal urban processes and, we hope, contributes to a more sophisticated and detailed understanding of how regulation works in its political-economic context.

Assessing Regulation

Past research surely suggests that whatever community and environmental benefits are created with growth controls are not attained in a direct trade-off for a dysfunctional economy. But does all of this huffing and puffing about growth—policy making, voting, regulating, deregulating, and project review—come down to nothing at all? Obviously we need sustained research on the effects of the last generation's regulatory efforts on land use as well as other fronts. Beyond the issues of whether growth control ruins economies or not, we need to understand precisely how new sets of building rules work and in what ways they help or hinder local community goals—economic and otherwise. The overall amount of new development may not significantly change, but the character of growth may shift in important ways, with controls actually hindering some kinds of development but stimulating others. Or, it may be that controls really do temper the pace of growth in some communities, but that other policies that are thought of as growth controls accelerate development in different places, or in the same community at a different time.

By focusing our attention on a specific set of case studies drawn from Southern California, we are able to address these questions using a variety of methods and encompassing a diverse range of control measures. We hope to transcend prior work, however valuable, that has relied on individual case studies (e.g., DeLeon, 1992b; Pincetl, 1992), on quantitative comparisons, as well as more anecdotal comparative work. In presenting our findings, we hope to lend insight and build the substance of broader public debates about how to mix private market mechanisms and public controls effectively. Were "private" development markets simply overridden by public controls, or was there a more "balanced" interaction of public and private? What would such a balance look like? What was the impact on local economies in these particular cases? Are they less efficient, or can cities thrive economically under "greener" regimes? Did communities grow through the "loopholes" in public policy, or was development managed through deliberate public trade-offs? Judging from the concrete results of local regulation, did local controls create a new era of citizen power over long-standing economic interests? Did the growth machine grind to a halt or just change gears? Did developers and their allies dig deeper into their "systemic power"—based on ideological advantages and institutional resources (Stone, 1981; Alford and Friedland, 1985)—to co-opt or otherwise derail their opponents' efforts?

If citizen resistance results, in fact, in effective and economically viable controls, this suggests the potential benefits of other political innovations that might seem to "interfere" with markets. Almost any effort to regulate on behalf of quality of life and environment is likely to be attacked as countermanding the wisdom of the market. Environmentalism at any level must respond to these charges: The successes of urban environmentalists may offer a model.

Plan of the Book

Now that we have presented the questions that drive our research, we will go on (in Chapter 2) to describe our study sites and the actual content of their growth control measures. Chapter 3 presents (in quantified terms) the impacts of these measures on building activity levels across our sites; that is, it answers the question of whether or not growth was "stopped." We then go on, in Chapter 4, to explain how developers and other growth interests function under growth control; we consider their strategies to overcome the impediments placed in their path—in other words, we look at how cities grow under growth control. In Chapter 5, we simulate a developer's efforts to build actual projects across our sites and thus begin the work of laying out the ways new regulations have indeed affected the content of growth. Chapter 6 continues this work of specifying the effects of growth control by analyzing the kind of growth localities experienced in relation to the kinds of development their civic traditions and planning policies tried to foster. Finally, we use Chapter 7 to sum up by assessing wins and losses, pointing to the ways intervention in land markets—and by implication other spheres as well—can improve the quality of life without sacrificing economic vitality, even in the narrow sense of the term.

Notes

1. Economists have often studied these "externalities" by proxy, assuming that all the pertinent negative spillovers and positive amenities are built into housing prices (see Fischel, 1990; Mark and Goldberg, 1981; Li and Brown, 1980).

2. Such influential environmentalists as Donella Meadows and her colleagues (1992) make a distinction between growth (which means getting quantitatively bigger) and development (which means getting qualitatively better by building local capacities). Although this is a valuable analytic distinction, in this book we use development in the more vernacular sense of what real estate developers do.

3. For evidence that low-income people bear the biggest pollution brunt, see Berry, 1977; Buttel and Flinn, 1978; Bullard, 1990, 1993, 1994.

4. "'Business' Is New Buzzword for Candidates" *Los Angeles Times,* May 23, 1994, A3.

5. For more detail and related treatments, see Molotch, 1976; Logan and Molotch, 1987; Feagin, 1983; Krannich and Humphries, 1983; Lyon et al., 1981; Domhoff, 1983; Swanstrom, 1985; Elkin, 1987; Whitt, 1982; Vogel and Swanson, 1989; Vogel, 1992; Jonas and Wilson, forthcoming.

6. Our simplification of the term "regime" does not conform precisely to traditional usage. See Elkin, 1987; Stone, 1989 for more definitive and elaborated statements.

7. See Eisinger's (1988: 152) summary discussion.

8. We first saw the phrase used by Brecher and Costello (1994), but the idea that governments are under pressure to price themselves below an appropriate level has been around for a number of years among economists, sociologists, and political scientists (see Cumberland, 1981; Molotch, 1976).

9. The classic studies are by Floyd Hunter (1953) and Robert Dahl (1961).

2

Sites

Learning from Southern California

Within the region of Southern California, so famed for sprawling growth, we focus on a diverse array of localities and a variety of measures aimed at growth control. We have no illusion that these localities and their regulations provide either random samples or "typical instances" from a nationwide perspective. Instead, our sites provide a test case of growth control with implications for other localities throughout the country. The sites in question are in a regional setting with both robust growth and widespread resistance to growth—apparently a twin dynamic. Research indicates that places experiencing rapid growth are more likely to turn to growth controls (Baldassare and Protash, 1982).[1] Southern California tends to be the "poster child" for environmentalists and business leaders alike, representing, respectively, the epitome of irresponsible development and the "excesses" of local regulation. We reason that if controls · over urban development have had important impacts anywhere, they would likely show in these settings.

Because of their relatively long experience with growth controls, our study sites should show some sophistication in designing statutes and enforcing their implementation. Controls should have been in place long enough to have measurable impacts. In contrast, new and tentative regulations would have undetectable effects. If little has changed in our study sites, the impact in many other communities and regions is probably even smaller; either their controls are minimal in real terms, or, given the lack of growth in their region, they were not needed in the first place. And the impacts of growth control that we observe at our sites may suggest future possibilities for communities that are just beginning to manage growth.

Studying the actual practices of growth regulations in California has a research advantage specific to the special qualities of the state. Large and culturally distinctive, California often gives momentum to broader national trends—as has happened with air quality standards, coastal pro-tections, and, emerging from different parts of the political spectrum, voter initiatives against affirmative action, and the taxpayers' revolt that yielded Proposition 13.

In the case of environmental issues, there is a firm economic basis for the "Californization effect." The vast California consumer market induces manufacturers to develop technologies to meet state environmental stan-dards and consumption patterns—in products like cars, construction ma-terials, pollution scrubbers, leaf blowers, aerosol sprays, waste systems, plumbing fixtures, and furniture—more readily than they would for other states or world regions. Sometimes manufacturers change their en-tire production just to accommodate the California market. Once this ac-commodation occurs, other states, even nations, can follow with similar standards more easily and ride the wake of California's rules because manufacturer's have already made the technological investment to pro-duce environmentally superior products. In this way, as well as in the force of its cultural and political winds, California's environmental poli-tics and patterns of life influence global economies, tending to set de facto standards. As we will show, growth controls sometimes involve effects, direct and indirect, on technologies that have the potential to impact oth-ers' modes of life.

It is a cliché to identify Southern California as a cultural forerunner, a basis for imitation because of its glamour and élan. But in regard to land use issues, this emulation has been much documented. The California bungalow of the 1920s and 1930s (Gebhard and Winter, 1977: 18), the "ranch house" of the 1950s and 1960s, as well as other residential and commercial forms (the fast food of McDonald's and many chains) began in Southern California—making it the most influential force in domestic architecture and "the center of contemporary restaurant innovation" (Pillsbury, 1990), with all that implies for resource use. The freeway and the hegemony of the American car probably owe more—economically, stylistically, and environmentally—to the Southern California landscape than to any other region. Thus, the stakes of what goes on in "local Cali-fornia" land use, as with most everything else, are high.

We could have used other parts of the United States in addition to Southern California, perhaps developing a broader comparative frame. But an analytic advantage of staying within a single region is that many of the economic, historical, and cultural factors are constant. For exam-ple, Southern California's economy, as measured by unemployment, wage structures, and property values, tends to rise and fall as a single unit (Levy, 1994). Hence differences that emerge among our communities over time cannot be attributed to divergent regional economic trends that would have to be considered were our cases drawn nationally.

Within the limits of the U.S. Constitution, state lawmakers have a great deal of leeway in setting the parameters for how local governments regulate land use. In practice, many states adopted the Standard Zoning Enabling Act of 1926 and the State Planning Enabling Act of 1927, which were developed by an advisory committee on zoning within the U.S. Department of Commerce (Platt, 1996: 234). More recently, the Model Land Development Code of the American Land Institute provided a template for updating enabling legislation, including provisions that some states used to adopt statewide land-use controls (Platt, 1996: 349). California has not gone as far in this direction as the exceptional cases of Hawaii, Oregon, Vermont, Maine, and Florida; but lawmakers have created a broad array of tools, ranging from coastal protections, to agricultural preserves, to environmental review to deal with land use and growth management (Platt, 1996: 365). Within our California study sites we encounter a single framework of state law, yet the foundations of this framework are enough in the national mainstream that we can provide analytic benchmarks for other parts of the country.

Selecting Study Sites

Restricting our study sites to the Southern California region was also practical. We could not settle for off-the-shelf databases; close and frequent on-site observations were needed to understand what an environmental control really meant and how it worked as implemented. Our work thus involved numerous site visits to carry out interviews, gather data, inspect archive materials, visit actual building projects, cross-check information, and attend dozens of public meetings. The two of us were able to carry out the work directly at every site with the support of several able assistants performing specific tasks. A broader geographical scope would have required more researchers and the kind of "hired-hand" coordination difficulties that create special challenges for qualitative work. In any event, employing staff and paying for air fares and extensive hotel stays would simply have broken our small research budget.

In selecting our study sites, we worked our way south from our University of California, Santa Barbara base, gathering up localities until our analytic needs were met. This method yielded three study areas, two of which contain an amalgam of distinct towns and cities that entered the case base in their own right. Our first study area is the Santa Barbara South Coast, consisting, in turn, of two incorporated cities and an array of unincorporated and quite diverse communities. The second study area is an even more varied set of communities that make up western Riverside County. Our third study area is the single city of Santa Monica, part and parcel of the surrounding city of Los Angeles in every way but its legal independence. Our study time frame is 1970–1990, the years that approximate the country's great environmental awakening and the era when growth-control measures were enacted. Besides starting early enough to catch later effects, this time frame gives us some opportunity to follow controls in a given locality as they work themselves out sequentially. In short, we have a variety of times, a variety of localities, and—as we will show—a variety of regulations upon which to build comparisons.

In terms of standard socio-demographic "predictors" of environmentalist sentiment, our sites appear to be a rough cross-section of California cities and towns. Some evidence indicates high-income and high-education populations are the most receptive to environmental sentiments and policy enactment, with working-class groups less oriented toward such issues (Dunlap, 1975; Lauber, 1978; Dowall, 1980; Protash and Baldassare, 1983; Clark and Goetz, 1994; but see Logan and Zhou, 1990). Our study encompasses places representing, if not the extremes, at least a reasonably wide socio-demographic variation (see Table 2.1 for the basic descriptions).

Identifying Growth Controls

Our list of what constitutes growth control in each locality includes measures that are not always put under that formal rubric. We relied on local understandings (using methods to be described) of the purpose and impact of a particular measure. Utility hookup bans, for example, are sometimes on the list and sometimes not, based on local perceptions. In some cases, we learned from informants on both sides that such moratoria were used to curtail development—even though the pro-growth activists were the only ones interested in proclaiming this publicly.

We gathered this information in the course of interviews (lasting as long as three hours, almost never less than one) that were designed to learn not only which controls were perceived to be most important but also how these controls impacted the community and how the implementation process worked out. We also probed for relevant historical information, including how the local political system had evolved and its implications for enforcement of restrictions. We examined specific projects and interviewed those responsible for their construction and oversight. In all our study areas, we interviewed planning officials (usually including the planning director), the mayor, and prominent developers. In each study area and in virtually all towns and cities, we also interviewed environmentalists, news reporters, and political activists on both sides of the development divide. In all, we interviewed ninety-three people. The largest number (forty-eight) came from the Riverside area because of the larger number of separate municipalities in this study area.

TABLE 2.1 Profile of Major Cities

	Santa Barbara			Santa Monica			Riverside		
	1970	1980	1990	1970	1980	1990	1970	1980	1990
Population	70,211	74,414	85,571	88,289	88,314	86,905	139,769	170,876	226,505
White	95%	86%	78%	93%	87%	83%	93%	81%	71%
Hispanic	21%	22%	31%	12%	13%	14%	13%	16%	25%
College Educated (over 25)[a]	18%	28%	33%	20%	34%	43%	17%	19%	11%
Median Family Income	$9,505	$20,285	$40,912	$10,787	$22,262	$51,085	$10,651	$21,075	$40,054
Families in Poverty[b]	9.20%	6.30%	7.84%	7.20%	6.60%	5.72%	8.30%	8.70%	8.44%
Housing Units	29,566	32,509	36,226	41,606	43,912	47,753	45,867	60,964	80,240
% Owner Occupied	45%	42%	40%	23%	22%	26%	62%	61%	53%
Median House Value	$25,044	$130,800	$345,800	$35,824	$189,800	$500,001	$19,643	$68,000	$134,400
Median Gross Rent	$133	$317	$715	$142	$319	$532	$120	$284	$575

SOURCES: Riverside County, 1970–1989; Santa Barbara County, 1970–1989; U.S. Census, 1970–1990; and U.S. Census, 1972, 1983, 1994.

[a] 1980 figures are for % with 16 or more years of education

[b] 1970 figures are families below low-income level

The remaining interviews were divided fairly evenly between Santa Barbara (twenty-five) and Santa Monica (twenty). These numbers do not include call-backs for information checks and briefer conversations (at least equal in number) that took place with assistants and others who were able to provide specific forms of information (dates of projects, meanings of specific conditions, etc.). We used semi-structured interviews tailored to the position of informants and to the kind of research issue for which their knowledge was most relevant. At each step, we asked informants to identify other key people involved with local growth and development issues, building our information sources in snowball fashion. With the permission of the respondents, we took extensive notes during the interviews, at times pausing to transcribe direct quotes. Afterward, we wrote up summaries, usually within a day, and catalogued interviews by the respondent's name, role, and study area. To assure confidentiality, we assigned each interview a number for reference purposes.

Our interviews were, in turn, supplemented by our scrutiny of documents, including the lists of conditions imposed on actual projects, the content of city and county plans, and records of regulations assembled by others. We subscribed to daily newspapers in all three areas and maintained clipping files on development issues generally as well as on specific projects.

Some people have made much of the difference between the terms "growth control" and "growth management," arguing that they are not the same thing. Hence a study of one would be different from a study of the other. Growth control, under this understanding, implies sharp limits, whereas growth management merely indicates mitigation of growth consequences.[2] But this distinction is less obvious in actual practice, as our findings will show. Legislation that looks like "control" may turn out to be more like "management"; growth is indeed allowed but only after the developer sweetens the deal with public benefits. Similarly, measures officially classified as "management" in fact "control" when administration is strict and backed by an anti-development political regime. The terms are themselves tools of local rhetoric as much as policy intent. Environmentalist policy makers sometimes use "manage" because they fear the word "control" will put people off; and developer groups sometimes dub mild managing devices "control" in order to paint their opponents as extreme. Our rule was to include measures that both sides seemed to think had the strongest impact, regardless of whether the measures were deemed "management" or "control."

Our study sites exhibit the kinds of differences thought to create favorable conditions for effective growth controls, middling conditions, and unfavorable conditions for effective regulation, in terms of socio-economic standing, geographic circumstance, and political-cultural histories. These

FIGURE 2.1 Map of Study Sites

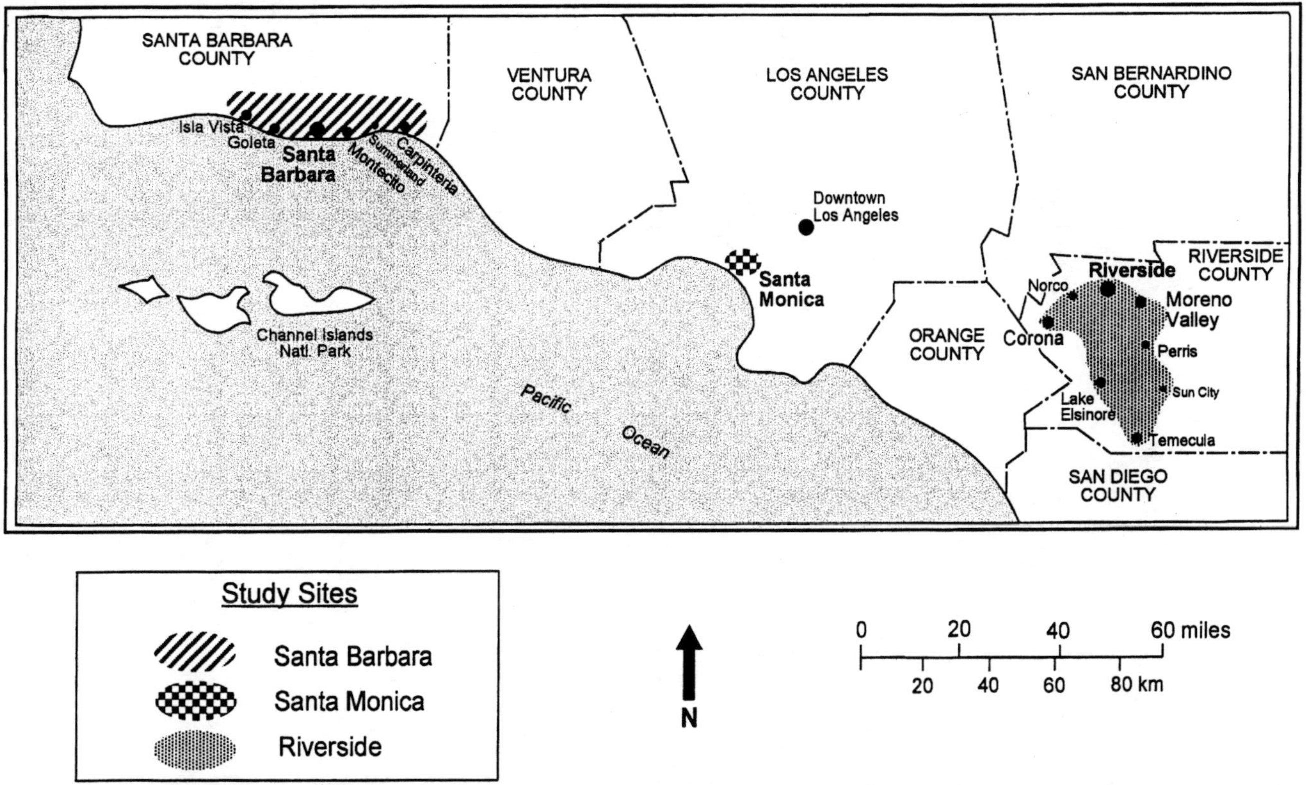

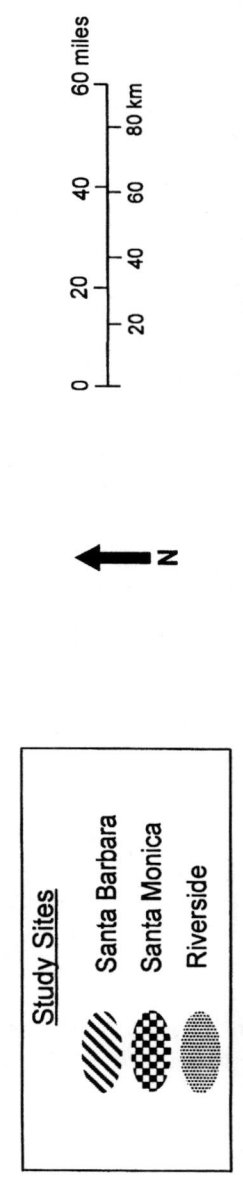

SOURCE: University of Colorado at Colorado Springs, Sheryl Giglia, cartographer.

places also differ in the type and apparent intensity of their controls, as we now elaborate through a description of each study area and its constituent locales.

Santa Barbara South Coast

Background

The Santa Barbara South Coast study area is optimally suited for growth control implementation. In geographic terms (see Figure 2.1), it is rather self-contained as a narrow shelf of land between the Pacific Ocean and the Santa Ynez Mountains. This portion of the Pacific coastline runs east and west and is located just beyond easy reach of Los Angeles (about a two-hour drive). Comparatively, it had more than the usual capacity to determine its own future, environmentally as well as in other ways, without significant spillover complications arising from inter-local commutes or neighborhood effects from adjoining zones.[3]

The Santa Barbara area was a distinctive settlement even before the Spanish conquest. Along with the Gabrielino people to the south, the indigenous Chumash people have been called "the wealthiest, most populous, and most powerful ethnic nationality in aboriginal southern California" (Bean and Smith, 1982: 538; see also Kroeber, 1925: 550, all as cited in McEvoy, 1986: 27, 28), with an advanced state of craft and artistic work (as evidenced in surviving basketry and cave paintings). In the modern epoch, special wealth and artistic assets mark the locale, and the area is sustained economically by agriculture, research/development, retirement, and tourism. The city of *Santa Barbara* (population about 85,000) has been the center for commerce, finance, government, and media. Although legendary for the riches of its upper crust, Santa Barbara is diverse: 30 percent of the core city population is Latino, and the median family income (within the city) is below the overall state level (although above medians of Californians living in other cities in the state).

To the west, Goleta is an unincorporated area of the county adjacent to the University of California, Santa Barbara's student dormitories and apartment blocks, otherwise built of tract homes and an array of research and development and light industrial activities. To the east is the exclusive suburb of Montecito, the world-class residential zone of impressive homes and estates that is actually the primary source of the locale's reputation for splendid affluence. Further east is the largely residential unincorporated beach community of Summerland.

The only other incorporated city along the South Coast is Carpintería (population 13,747); a substantial portion of its population is working class and/or Latino. In years past, it served as the commercial sub-center for

Santa Barbara city view

nearby farms and was home to working people employed elsewhere in the local region. Only at the end of our study period (in 1990) was its traditionally pro-growth city council replaced by an environmentalist majority.

The city of Santa Barbara has long been the center of planning and land-use politics—not only for the region but in some respects for the state and the nation. Some say the 1969 Santa Barbara oil spill not only stimulated local environmental protection (and the enactment of what an oil official called "shamefully excessive" demands; see Horner, 1990: 4), but sparked the global environmental movement (Easton, 1972). In fact, such concerns go back at least as far as the 1925 earthquake and fire; the city was rebuilt under the supervision of the country's first architectural review board. A homegrown version of the Andalusian Spanish architectural style (which became the city's trademark and inspired its many grand and noteworthy official buildings and commercial structures) was explicitly promoted by local government under the influence of citizen activists (similar to the invention of the "Santa Fe" style, see Wilson 1997). At the county level, Santa Barbara also became the first in California to require architectural review (Gebhard 1990: 7).

Historically, city government leaders were drawn mostly from "Main Street" business groups, typical of small-town America. But on certain issues and projects, the official leadership deferred to a civic-minded elite who used their prestige and purses to block development—for example, by buying land for public open space along the beachfront until the city could be induced to pay for it. Although this state of affairs was no doubt tense at times, there was a modus vivendi in which citizens' groups "watched over" a growth-oriented local government and controlled, at least in terms of the aesthetics, how development occurred. This proved a workable truce as long as growth pressures remained moderate.

These conditions began changing in the late 1960s, particularly with the dramatic growth of the University of California's new campus in Goleta (officially the University of California, Santa Barbara). County voters set a new precedent in 1970 by overturning their elected officials' approval of a development—a 1,535-unit "leapfrog" luxury housing project north of Goleta by the famous "El Capitan" surfing beach (Graves and Simon, 1980). In a similar 1972 action, Santa Barbara city voters stopped an eight-story residential condo "tower," later city officials wrote a permanent citywide four-story height limit into the city charter. "In all ten Santa Barbara-area referenda involving environment and growth issues between 1967 and 1983, the environmentalist position prevailed" (Sollen, 1983).

Certain elements of the community—notably the Citizen's Planning Association, with its roots extending to the post-quake rebuilding period—had long promoted "good planning." The planning advocates included retirees and the wealthy, bolstered by two other groups: (1) those

Santa Barbara beachfront on Cabrillo Avenue

who were buffered from the local growth economy by working for large, non-local public and private organizations—particularly the university and the high tech industries; (2) those whose business was Santa Barbara's scenic character, namely tourism (this group provided less consistent support).

By 1990, the most important elected bodies at both city and county levels were controlled by people who won office as environmentalists, some of them as long as twenty years earlier. The environmentalist rise to power was not a steady progression but a back and forth contest for political position, especially at the county level. The environmentalist dominance achieved by 1990 was, in fact, short-lived, as evidenced by elections after our study period that returned control of some jurisdictions to the pro-growth camp. But the array of elected posts under environmentalist control at so many levels (city, county, water boards, and other special districts) and for such an extensive period (most of the twenty years) may have been unique in America.[4]

Environmental groups in the city of Santa Barbara, in particular, institutionalized their presence to a high degree. The Citizen's Planning Association (with a countywide reach from its city base) maintained a downtown headquarters and employed a planner to assist with lobbying and development monitoring. Network, a left-oriented citizens' action lobby, sustained an office at times and usually employed at least a part-time staff person. Other such organizations came and went, a later example being the Santa Barbara Alliance, which raised money for political candidates and monitored the development process on a daily basis. Groups such as the Citizens for Goleta Valley proved equally effective, though they were fully volunteer. The Environmental Defense Center (with four full-time staff lawyers, plus paid support staff, interns, and volunteers) waged litigation wars on land-use issues. The Community Environmental Council, founded in 1969 as an environmental action and research center, developed a wide array of programs with a paid staff that grew eventually to sixty persons (including recycling center workers). The council came to have its own showplace building funded by a local philanthropist.

Beyond these general-purpose and locally autonomous groups, Santa Barbara had extremely active chapters of national associations, especially those with strong environmental concerns (e.g., the Sierra Club, the League of Women Voters). The local Audubon Society, at the time of the 1969 Santa Barbara oil spill, had a remarkably high number of members, with 750 in the local chapter (Molotch, Freudenburg, and Paulsen, 1999). There were also groups oriented toward more particular interests, like maintenance of horse trails in the adjacent national forests, support for bicycle transportation, surfer's rights to unpolluted ocean waters, and ongoing efforts to block expansion of the offshore oil industry (e.g., Get

Oil Out!). Still other groups sprang up in response to specific projects. These groups often functioned in coalition, and Santa Barbara was regarded by developers and environmentalists alike as a strong bastion of environmentalist sentiment and organization. By the mid-1970s the region had become what the *Wall Street Journal* called "a major proving ground for various measures that have held back residential, commercial and industrial development" (Hill, 1978: 1, 27).

Growth Controls: Santa Barbara South Coast

Across the variety of localities making up the South Coast, a diverse array of controls had been enacted over time, through voters' initiatives, board and council actions, commission implementations, and administrative policies (Appendix B lists controls for each locality in chronological order). Almost every control was controversial, usually with intense public debate and often with great acrimony.

As is true throughout the arid West, water is the mother's milk of Southern California's growth. The first move toward growth control on the South Coast was the 1972 Goleta Water Board decision to grant no new water hookups in what had been the county's fastest growing and largest suburban area. Voters ratified this decision by popular referendum and installed an environmentalist majority on the water board in 1973. In 1979, county voters elected not to tap California's state water system (the only California county to do so). Two smaller water districts (Montecito and Summerland) imposed their own (less stringent) moratoria and rationing limits (in 1973 and 1974, respectively).

In 1974, thinking ambitiously about future policies, the city of Santa Barbara commissioned a group of local academics and environmentalist volunteers to determine how different levels of future population growth would impact the city. In effect this was a giant environmental impact report on the effects various growth scenarios would have on all aspects of fiscal health and city services. Based on the three-volume report (Appelbaum, et al., 1976), the city council selected the figure of 85,000 as the city's final capacity (10,000 more than were present at the time) and downzoned virtually all multi-family residential areas to approximately half their prior allowable densities. In the face of legal challenge, voters then (in 1977) endorsed both the population limit and the downzoning. The same measure advised that all changes to the city's general plan be submitted to a vote of the people. Taking it a step further, a 1982 vote added to the city charter the principle that the city was to "live within local resources"—a not so veiled proviso intended to discourage any future commitment to the state water project. Still another provision required that five council votes would be needed for any future rezoning (a "super-majority" of the seven council members).

In a move that seems incredible given the discourse that followed in the 1990s, the city acted to "balance" its housing and jobs by decreasing the number of jobs it would potentially attract. Council action and a subsequent citywide vote (Measure "E" in 1989) sharply limited future commercial building (retail, office, and manufacturing) to a total of three million square feet over the following twenty years. Commercial buildings could have a floor area—"footprint"—equal to only one quarter of the land upon which they were built; small additions (under 3,000 square feet) could be made to existing structures, but any larger projects had to compete for the remaining commercial space not allocated to projects in the pipeline (approved or pending) or to "community priorities." Since two and one-half million square feet was already set aside for such uses, only one-half million square feet remained to accommodate new proposals (a figure that translates roughly as a two-department-store shopping mall available for all future commercial functions including office buildings, manufacturing, and so forth).

On the water supply front, in 1986 a long drought caused the city to disallow commercial developments that required any water beyond a parcel's historic use, with subsequent restrictions requiring net water savings for all projects. The city limited new residential hookups to about fifty per year (allocated on a lottery basis).

More or less keeping pace with city residents, in 1978 county voters approved a 0.9 percent annual population growth limit for the unincorporated areas of the Santa Barbara South Coast. The 1989 county growth management ordinance limited the Goleta area to 280 new housing units and 80,000 square feet of commercial development each year (a very small increment, given the approximate 75,000 person population base).

The small city of Carpintería remained a relatively unregulated zone throughout this period, with growth resistance taking hold only in 1990, when voters rejected city plans to establish a redevelopment agency that was perceived as a tool for growth promotion. Under the rallying cry to "save Carpintería's small town character," a large beachfront, mixed-use complex was turned down by the freshly elected environmentalist city council.

Santa Monica

Background

Santa Monica approximates Santa Barbara in population size (86,905) but was our middle case in terms of the potential for effective growth control implementation. Compared to Santa Barbara, its options were strongly constrained by its location at the seaside edge of the great L.A. metropolis.

Pacific Coast Highway in Santa Monica

Beyond the city line to the north are affluent L.A. suburbs; to the east and south, Santa Monica is woven tightly into the West Side of Los Angeles, its borders indistinguishable from adjacent L.A. What goes on in the giant metropolis feeds in as traffic congestion, housing pressure, homelessness, crime, and housing price inflation.

It was not always so. Santa Monica began as one of the first Southern California coastal resort towns; travelers making a two-day wagon trip from downtown Los Angeles would camp on its broad beach (Banham, 1971). It grew along with the L.A. region during and just after World War II as military production (Douglas Aircraft in particular) became a center for local employment. The population was heavily working class, living in modest bungalows. The final link with the metropolis came with completion of the Santa Monica Freeway in 1966.

As integration with the rest of L.A. proceeded, Santa Monica's population became more socially heterogeneous, with African-Americans and Latinos taking up residence, particularly in the southern portion. The summer beach homes and apartments in the Ocean Park area attracted a more settled community than before, at once bohemian and politically progressive (Shearer, 1982). Renters became an ever larger and more vocal proportion of the city's population, reaching 78 percent in 1980 (Clavel, 1986: 141). The city also became home to high-rise residential, hotel, and office structures, with a particularly strong building boom in the 1984–1989 period. Pressures mounted for a massive intensification of existing land uses and the gentrification of much of the housing stock. Sound and substantial single-family homes were bought (at prices in the one-half million dollar range) as "tear downs" and replaced by much larger single-family structures crowded onto ordinary sized city lots. Similarly, low-rise apartment blocks and commercial structures became marketable for the value of their land alone. Santa Monica's relatively clean air, its proximity to beach recreation, and its easy access to L.A.'s important business and cultural centers together sparked intense growth pressure (Fulton, 1985: 63).

Through most of the 1960s, civic life in Santa Monica was controlled by the hometown elite of the Chamber of Commerce and the service clubs (Condelli, 1980). These groups were subsequently challenged in successive waves. The first sign of challenge came in response to a 1979 city council plan to tear down the Santa Monica pier; the grand vision included an ocean causeway arcing north through the Pacific Ocean to link the city to Malibu, with an artificial island for high-rise development built on bay landfill as an additional feature. Citizen protest stopped the project, with even the normally pro-growth local newspaper, the *Outlook*, in opposition. Several of the liberal Republicans who led the fight to save the pier were subsequently elected to the city council. Although not opposed to the underlying premises of growth, they rejected this project as at best impractical and at worst disastrous for the city's character.

The first effective challenge to growth itself was a citizen lawsuit to block public financing of a proposed downtown shopping mall ("Santa Monica Place") on the grounds that the project would not benefit neighborhoods. Though financing for the project was initially approved by voters in 1975, the city ended up settling with the plaintiffs in 1978. The settlement provided that a portion of the project's parking revenues be channeled to low-income housing rehabilitation, pocket parks, recycling, child care, and a women's center. This development agreement became a model for how the city would later negotiate with developers.[5]

By late 1978, several renter groups banded together to form the Santa Monica Renter's Rights group (SMRR). SMRR won rent control with the city's highest voter turnout in its history. Over the next few years, SMRR gained control of the newly established rent control board and the city council (control of the latter was to be lost for a five-year interregnum). Because of both its strong regulation over rental markets and the level of citizen activism across many fronts, including land use intervention, Santa Monica became one of California's most prominent examples of proactive local government. Although rent control was the electoral issue that first transformed Santa Monica politics, it was delivered within a broader challenge to the power of landlords and developers over civic affairs (Carnoy and Shearer, 1980; Shearer, 1982). The liberal and radical leadership gave the city a reputation for severely curtailing private property rights in favor of citizen and neighborhood interests (conservative groups called it, derisively, the "People's Republic of Santa Monica"). The city was the laboratory, activist base, and main constituency for the Tom Hayden and Jane Fonda's Citizen's for Economic Democracy (CED), as well as the district represented by Hayden during his career in the California State Legislature. Compared to activists in Santa Barbara, Santa Monicans were concerned not only with preserving the physical environment but also with a strong commitment to a social justice agenda (Heskin, 1983; Carnoy and Shearer, 1980; Capek and Gilderbloom, 1992).

Growth Controls: Santa Monica

When the SMRR rent-control challengers gained a council majority in 1981, they stopped issuing construction permits immediately, even for projects already approved. During a six-month permit moratorium, new city leaders assessed how much development should be allowed and what social benefits should come from it. Projects "stuck in the pipeline" were allowed to proceed only after new terms were negotiated, with extra benefits for the city under case-by-case "developer agreements."

Santa Monica Place Mall

During the initial moratorium, the council—along with citizen commit-tees—devised stricter reviews, new development guidelines, and more development fees—including fees for impacts on traffic and schools, and to provide affordable housing. The city negotiated exactions on a project-by-project basis and, later on, with the guidance of revised land-use stan-dards developed in the comprehensive plan of 1984.

In May 1989, a newly installed SMRR majority on the city council im-posed another permit moratorium to stall the boom in commercial de-velopment pending development of a growth management strategy. This action came only nine months after the city adopted a new comprehen-sive zoning ordinance, which reduced total allowable commercial/indus-trial development in the city by two-thirds. Meanwhile, growth critics were able to block several large-scale new developments, even ones that promised direct revenues to the city and special social benefits, such as redevelopment of the city airport and construction of a beachfront hotel on the publicly owned site of an existing private club. Santa Monica's tight restrictions grew tighter still.

The Riverside Area

Western Riverside County, the most far-flung and fastest growing of our study sites, represents the stereotypically least conducive scenario for ef-fective growth control. Its proximity to the spreading metropolis pulls development its way, and local growth boosters refer grandly to this in-terior region east of Los Angeles as the "Inland Empire." The area's heav-ily working-class population base and weak organizational infrastruc-ture also imply more tentative growth resistance.

Background

Our study area includes not only the city of Riverside but those county portions that stretch from the L.A. County border on the west, to Orange County on the south, to adjacent San Bernadino County on the north, and forty miles east into Riverside County(roughly, our area is a forty-mile-by-forty-mile square). Hemmed by mountains on all sides (some rising above 10,000 feet), the enclosed valleys are studded with rocky outcrops and edged by broken badlands that carve them into smaller set-tlement pockets of distinct natural character. The fact that anyone could consider all of this part of the same urban region reflects the magnitude of development pressures spilling into Riverside from the coastal areas.

Riverside's natural environment (hotter in summer, colder in winter, drier year round, and without access to the sea) was less hospitable than Santa Barbara's to people who lived without extensive public water

Inland Empire National Bank sign

works and interstate freeways. At the time of European arrival, what is now Riverside County had a population of 4,000–5,000 Cahuilla people (Stanton, 1989b). The Spaniards and Mexicans made few land grants, and settlement remained sparse under their rule. Interest in the area was sparked only during the railroad era of the late nineteenth century. The city of Riverside was established in 1870 by educated abolitionists, mostly from Iowa and Michigan (Patterson, 1971), with other investment groups buying land tracts nearby. These pioneers introduced irrigation systems and the orange groves that were later to cover large expanses. Early city leaders did take an interest in formal planning, hiring Charles H. Cheney (following his work in Santa Barbara) in 1927 to create a master plan for a graceful city of broad, tree-lined avenues and a high level of amenities (Patterson, 1971: 294).

Along with much of Southern California, the Riverside area grew steadily, though by 1950 it was still largely agricultural. In that year, less than 50,000 people lived in the city of Riverside, and Corona was the only other town with more than 10,000 (and just barely). It was not until the late 1970s and especially the 1980s that the larger Riverside area "took off" as part of the thrust of L.A. expansion. Developers constructed large tracts of inexpensive homes, a pattern still evident years later. The median sale price of a Riverside/San Bernadino house in 1989 was $129,920, compared with $222,842 in Los Angeles and $253,034 in Orange County (Kristof, 1989). Although the stereotype has been that Riverside growth comes from L.A. people "moving out" to find affordable housing, a 1989 poll by the *Riverside Press Enterprise* found that 68 percent of those surveyed were either longer-time residents (31 percent), had moved from other areas of the state (16 percent), or were from out of state (21 percent) (Stanton, 1989a). If there was a large "spillover" from the urbanized counties, it was probably not people moving out to Riverside but mostly a displacement of people that may have otherwise moved to Los Angeles or Orange counties.

All of western Riverside County experienced growth pressure, but the effects on each jurisdiction were somewhat different—as were the reactions of each. The county government, the jurisdiction that could reasonably give the great growth spurt an overall direction and control, was not generally activist, even by the standards of the earlier 1970s period. Most noticeably, there was little effort, or even awareness that it might be a worthy goal, to coordinate planning efforts with adjacent San Bernadino County, with which Riverside County shares important transportation corridors as well as common problems of rapid growth and air quality deterioration. Our interviews with local planners in both counties, people who by profession are usually highly committed to regional perspectives, revealed that even *they* thought about the two counties separately,

Teenagers in downtown Riverside

which implies that the cities of the region were left to lead efforts to control development.

The city of Riverside is the largest and most influential urban center of the Inland Empire, with a 1990 population approaching one-quarter million people. Its original citrus industry still visually dominates much of the landscape. Over the years, Riversiders fostered a reputation for a certain level of sophistication and urbanity, especially compared with adjacent rough-and-tumble San Bernadino just across the county line, which housed more of the railroad workers and other working-class people. The county of Riverside was created in part to break the city of Riverside away from San Bernadino (Stanton, 1989b). The city of Riverside became the home of the county government, and Riversiders continue to develop and support cultural resources, including an orchestra and a museum. The city led efforts to turn an agricultural research station into the University of California, Riverside. The campus is an impetus for local growth and a focal point for regionwide cultural activity.

Ringed by shopping malls both within its own boundaries and in adjacent jurisdictions, the downtown languished for several decades. Its grand old hotel, the Mission Inn (site of the Nixon honeymoon, as well as of other historic events) stood forlorn and deserted in the city center during our study period, despite large infusions of capital to revive it as a downtown cultural anchor. The city abounds in buildings of architectural interest, including pleasant residential districts of classic California bungalows and mission-style buildings.

Corona (population 76,095; fifteen miles southwest of Riverside on Highway 91) was founded in 1886 by its own core of speculators and citrus farmers in rivalry with the city of Riverside. It took about 100 years for the grand vision to begin to bear fruit. Adjacent to regional freeways, Corona was a fast growing boomtown of commuter-oriented residential tract development by the late 1980s; by 1987 it had surpassed the city of Riverside in the amount of annual commercial space approved for development. Toward the end of our study period, it was among the state's fastest growing cities, providing moderate- and low-cost homes and modern retail settings for a vast new population. By other lights, it is an ecologically wasteful prototype of residential and retail sprawl with no coherent cultural center, street life, or employment base.

Hemet (population 36,094; thirty-three miles southeast of Riverside and thirteen miles from the nearest freeway) is a retirement center and has grown steadily but slowly since the 1960s. It is also something of a satellite agricultural center, but it was dwarfed in the 1980s by the booming towns around it.

Lake Elsinore (population 18,285; fifteen miles south of Corona) survived for many years as a predominately working-class resort town by exploiting its natural reservoir. In the early 1900s a train line brought vacationers from Los Angeles to this "clean air" desert community. The lake that gives the town its name remains an attraction, bringing in jet skiers and ski boats. In 1988 the state completed the final link connecting the city into the L.A. freeway system,[6] putting Lake Elsinore within commute range (by regional standards) not only of Los Angeles but of Orange County and San Diego as well (fifty miles to the south). By the end of 1989, some 25,000 units of new housing were approved but not yet built within the city limits, potentially adding an estimated 75,000 to the 1989 population of 15,000. The impending residential boom promised to transform Lake Elsinore into another suburban bedroom community.

Norco (population 23,302; just north of Corona) was established in the 1960s as a bedroom suburb for horse enthusiasts. Consistent half-acre, low-density zoning includes riding trails that connect all houses; there are no curbs or street lights. The town has little land left for development within the city limits, but Norco officials had scant interest in annexing new territories (they even gave up a little land to Corona in one case). On the other hand, the city actively joined the commercial development game, promoting the Norco Auto Mall, with other potential projects in the wings.

Perris (population 21,460; seventeen miles southeast of Riverside) was known as the region's "welfare town." Among all the places in our study areas, it has the lowest median income and highest proportion of people receiving government assistance. Several recreational vehicle factories provide some jobs, as does nearby March Air Force Base. In the 1980s, residential growth began to take off, with the construction of lower-cost housing within the city limits. The town has tried to encourage commercial/industrial development as well.

Sun City (population 14,930; about five miles south of Perris) is a prototypic retirement community. Built by developer Del Webb after his hugely successful retirement Mecca of the same name in Arizona, unincorporated Sun City has a limited degree of self-governance through a county-approved Community Service Area (CSA) status. This allows the collection and use of fees for some services.

Moreno Valley (five miles east of Riverside on Highway 215) grew "overnight" from a population of 8,000 in 1980 to a city of almost 120,000 ten years later—drawing boomtown recognition all over the state. Initial growth was driven by huge residential developments approved by the county—some at extra high density under a county program to encourage "affordable" housing. After two unsuccessful tries, citizens' groups led the successful effort to incorporate the city of Moreno Valley in 1984—in part as a first step toward gaining some control over the rate of residential construction. Consisting of adjacent subdivisions, there is no visual, cultural, or business center to this city, nor any apparent physical relationship to the adjacent strips of commercial development.

Moreno Valley sprawl

Once incorporated, the town tried to improve a fiscal situation made desperate by so many tracts of modestly priced suburban houses. It battles for more commercial development, primarily to gain sales tax revenues (cities get a piece of the action in California; the rest goes to the state). Its major shopping complex went into head-to-head competition with another regional mall planned literally next door in the city of Riverside. As was common in the Inland Empire, two local economic development agencies are pitted against each other in seeking tenants and offering deals.

Temecula (population 27,099; nineteen miles south of Lake Elsinore) is in the southernmost part of our study area. Although incorporated only in 1989, it had some coherent identity prior to this time because it was originally a master-planned community rather than just a collection of subdivisions. The project was taken over as "Rancho California" by Kaiser Aluminum, which later sold the remaining vacant acreage to another large developer. Although the master plan includes a rather balanced mix of residential and commercial development, infrastructure and services are woefully inadequate. Under county building standards, the road system was only built to meet rural standards, resulting in intense traffic congestion once the population grew in the 1980s.

During our study period, the cities in the Riverside area varied from one another in many regards and encapsulated a number of development conditions found among growing localities throughout California. Only the city of Riverside had anything approaching a sturdy planning tradition. Across the region, at least compared to Santa Barbara and Santa Monica, government intervention—when it did take place—occurred in the context of developer pressure and in the absence of sustained citizen mobilization. Nevertheless, local governments throughout this area did create some form of growth control in virtually every jurisdiction.

Growth Controls: Riverside Area

The Riverside area controls were sufficiently numerous and well-known to "count" in the listings that hearten environmentalists and dismay developers. Rather than imposing aggregate growth caps, the most notorious (or appreciated) restrictions in the Riverside area generally protected special subareas and specific natural resources, although there were exceptions like the tight Norco horse zoning and the Sun City build-out.

In 1979, voters in the city of Riverside created an agricultural zone to preserve citrus groves within the city; the groves were increasingly appreciated not only as "open space" but for the nostalgic beauty they lent the palm-lined drives through which they were routinely viewed by residents. Housing in the citrus belt was held to one house per five acres (Proposition R)—a ratio that precludes tract-style housing but would not qualify as "agricultural" at all in some localities. The measure also decreased the density allowed in hillside developments. In 1987 voters tightened the restriction (Measure C), making it more difficult for a city council to dilute the earlier downzoning, and blocking a blue-ribbon group's recommendation to facilitate cluster development in the citrus zone. Sometimes favored by environmentalists because it encourages open space, cluster housing allows developers to combine allowable units over a larger tract into a single dense complex; it would have created, in effect, higher overall densities by lowering development costs and by allowing land difficult to develop to "count" as acreage. In this case, environmentalists would not accept the trade-offs.

An earlier precedent for citrus protection was the city of Corona's 1977 downzoning of 5,000 acres of citrus trees to preserve agricultural use. The "horsey town" of Norco followed Riverside in downzoning hillsides, requiring a full acre of land for every home site (Norco's founding half-acre minimum in the rest of town could be thought of as itself a strong form of growth control). In the southwestern part of the county, environmentalists raised funds to establish the Santa Rosa Plateau grasslands as a nature preserve.

Sewers were also a limiting factor. Beginning in 1982, the city of Riverside, though exempting annexations and redevelopment projects, restricted the number of new waste water hookups. In response to inadequate sewer and flood control capacity, Corona enacted a building moratorium in 1976, limiting residential permits to 450 per year. This was extended through 1977 when the city, while successfully tapping into regional sewer lines, adopted a point system for new projects to encourage more efficient land-use patterns.

When voters in the city of Riverside strengthened Proposition R in 1987, they also forbade annexing any more lands unless adequate infrastructure was available (failed growth management measures for the county and for Moreno Valley had included similar provisions). A general initiative did pass in the city of Hemet, requiring adequate infrastructure before development approval. Other cities have council-approved ordinances requiring advance adequate infrastructure, particularly for commercial and industrial developments. In Corona, minimum lot sizes were nearly doubled (to 7,200 square feet) by popular vote in 1986 because of the same kinds of infrastructure concerns.

Although not ordinarily considered an aspect of growth control, and indeed more usually motivated by the opposite goal (see Hoch, 1984), a number of city incorporations had environmental concern as a background motivation. Thus, the impetus for incorporating Moreno Valley was to ensure more stringent review of development standards. Temecula incorporated for similar reasons.

Among the most important development constraints in the region was one that came not from local policy but from a federal mandate implemented by the U.S. Fish and Game Department. The listing of the Steven's Kangaroo Rat as an endangered species put the brakes on a number of projects—including those proposed by local governments. Other federal agencies were involved in the protection of wetlands—the dried up gullies of the desert are also part of the waters of the United States.

The most ambitious effort to deal with all these problems, and more, was the countywide growth control initiative that went down to defeat at the polls in 1988. As is usual with such endeavors, the environmentalists were badly outspent, and defeat was urged by virtually every media outlet in the county. But the fact that 40 percent of the voters supported such a measure was motivation enough for county officials to try to "do something" about rapid growth. At the combined request of the local building industry association and the Sierra Club, a growth management element was developed for Riverside County's comprehensive plan. The final document contained a range of policies concerning growth (including some preservation measures), though it did not define how much growth should be allowed in the end or how quickly that growth should happen, delineating instead *where* growth should occur.[7] The document made the news as action on growth control and joins the inventories of lists of government measures taken to deal with development problems.

Growth Controls in Place: A Summary

Our review of enacted policies across the jurisdictions in our study areas indicates that communities took a wide variety of formal actions, which together appear as a more or less continuous stream of efforts to intervene in the development process during this time period. There are differences between localities, both in the degree and type of controls. Riverside area jurisdictions emphasized the preservation of certain zones and used utility hookup restrictions to deal with infrastructure shortages but mostly desisted from setting total holding capacities or limiting rates of growth. The city of Santa Barbara had a long-standing cap on residential build-out, with a more recent limit on commercial development. Santa Monica relied more heavily on project-by-project regulation as a means of control. The abundance of growth control techniques in place poses the possibility that overall development was stymied across our study area, the problem to which we now turn. Let's look at the numbers.

Notes

1. For more analysis of which localities tend to adopt growth controls, see Dowall, 1980; Glickfeld, Graymer, and Morrison, 1987; and Donovan and Nieman,

1992. Although these findings were important to us in choosing our sites, we grew increasingly concerned that their dependent variable "growth control" relied too heavily on formal government statements and rules rather than substantive implementation (see also Logan and Zhou, 1990).

2. For alternative discussions of this distinction see Gottdiener (1983).

3. Although the extent of commuting, especially from lower-housing-cost north county locations, increased during our study period, the significant distance of the city from adjoining urban areas gives it a level of autonomy (in regard, for example, to traffic planning and air quality maintenance) not found at our other sites. Median 1990 commute time was sixteen minutes for city of Santa Barbara residents (U.S. Census, 1994).

4. To draw a California contrast, Berkeley has also had strict environmental policies, but the surrounding territories have not; pro-growth regimes have largely controlled the Alameda County government.

5. Robert Myers, the legal aid lawyer that represented the plaintiffs, was later appointed city attorney.

6. Via Interstate 15 to the San Bernadino Freeway.

7. Orange County's experience in the late 1980s ran parallel to Riverside County's. In Orange County's case, a county-wide growth control ordinance (with real teeth), favored by a wide majority according to preelection surveys, was defeated after development interests put together a $2.5 million campaign against the ordinance; growth controllers supporting the ordinance spent $106,000. Nevertheless, developers led by the Irvine Ranch Company were frightened into generating certain mitigations, including means to finance road construction and initiatives for a program of open-space preservation to accompany future development (Pincetl, forthcoming).

3

Has Growth Been Stopped?
Not Much

To assess the impact of controls that our informants identified as most important in limiting development, we gathered building and growth data from eleven jurisdictions in our study areas. In this chapter, we first show general trends for commercial development, retail sales, and residential construction in our study areas. Then, focusing specifically on residential construction, we test statistically whether or not growth controls did, in fact, impact housing construction.

First Findings: Rates of Growth

In a nutshell, we found scant evidence that controls had much of an effect, particularly on the supply of new housing—the kind of construction most controls had been aimed at curtailing. First we will look at commercial growth rates, including retail sales, and then turn to housing.

Commercial Development: Boom

Even though most growth controls focus specifically on residential development, they are often portrayed as posing a general threat to the local economy. Increased regulation is said to create an "anti-business" climate that stifles growth. More specifically, increased housing costs, allegedly caused by new regulations, make it harder to attract workers to the area and to get them to stay.

Commercial development trends in these growth-managed communities did not conform with these representations. Instead, commercial building in each area increased dramatically over the twenty-year period. Comparing local trends with state and national patterns (Figure

FIGURE 3.1 Commercial Development Trends, 1970–1990

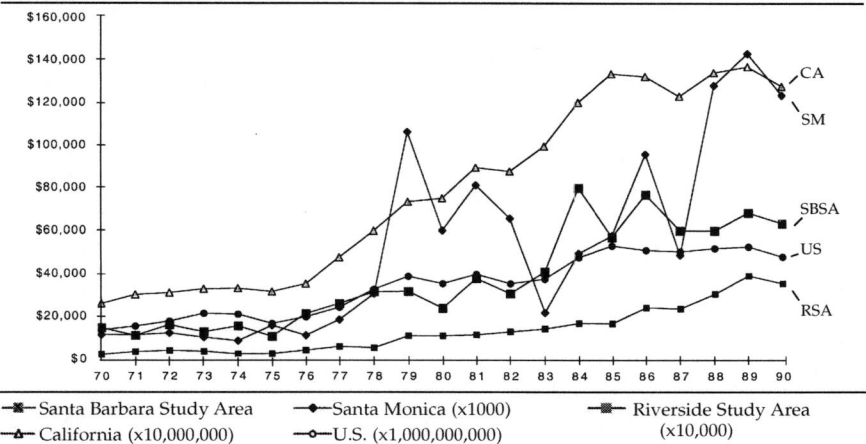

SOURCES: Construction Industry Research Board, 1971–1990; Riverside County, 1970–1989; Santa Barbara County, 1970–1989; U.S. Department of Commerce, 1970–1989.

3.1), we see that local investment, rather than responding to local growth controls, followed the broader economic dynamic.

In 1990, the value of new commercial development for the nation as a whole was 259 percent higher than the value of new development in 1970—much of this due to sheer inflation.[1] Commercial development in the Santa Barbara, Santa Monica, and Riverside areas rose significantly more than this over the period, though local development activity is more volatile from year to year. Some places far outpaced even the robust four-fold statewide increase in commercial development valuations during the two decades; for example, Santa Monica commercial development valuations grew 1,000 percent, and Riverside study area growth was close to double that. Commercial development was more moderate in the Santa Barbara study area, but valuations still increased a strong 341 percent in the same period—only somewhat less than the state overall increase and greater than the country's (see Appendix C, Commercial Valuation Data). Expanding municipalities and the proliferation of "greenfield" development in unincorporated suburban tracts drove the fast-paced growth in the Riverside study area. Overall, these areas, each with growth controls of some sort, kept up with, and even exceeded, broader trends; in particular, the Riverside and Santa Monica study areas (two very different business climates) did extraordinarily well.

TABLE 3.1 Percent Increase in Retail Sales, 1971–1989

	Percent of Increase
City of Santa Barbara	320
City of Santa Monica	368
City of Riverside	325
State of California	317

SOURCE: California Department of Revenue.

Growth in Retail Sales

Retail business, another measure of economic vitality, also thrived over the period. In the three largest cities in our study areas, retail sales (indexed by sales tax revenue) equaled or surpassed the statewide level, as indicated by Table 3.1, which shows the percent increase in per capita retail sales over the 1971–1989 period:

Growth rates in these central cities are especially impressive considering that traditional urban centers throughout the state and nation were losing retail activity to regional malls and suburban business strips. The vitality of the retail sector in the study sites would be even more striking had we been able to gather sales data for unincorporated areas, where most of the commercial growth almost certainly took place; building permit data showed that these unincorporated areas had the most new commercial construction, but we were unable to break out retail sales for unincorporated communities because of the way the state collects this data.

Growth in Residential Development

Significant numbers of new residences were also built over the two decades. Between 1970 and 1990, 17,298 housing units were approved in the Santa Barbara study area (adding 37 percent to the existing housing stock); 11,293 in Santa Monica (a 27 percent rise); and almost 180,000 in the Riverside study area (a 151 percent gain).

Although there are significant differences in building activity across our areas, the overall ebb and flow of residential building usually followed the broader state trend. Figure 3.2 compares residential construction in the study areas with the statewide and national pattern (see Appendix D for detailed residential building rates for all the study sites). In the early phase, around 1971, house building rates in all our study areas clustered closely with the state's overall rate but then fell off in Santa Monica and Santa Barbara, while rising well above the state in the Riverside study area. The fast and erratically growing Riverside study area is the deviant case in this regard, greatly outstripping state levels, especially in the 1980s.

FIGURE 3.2 Residential Building Rates, 1971–1990

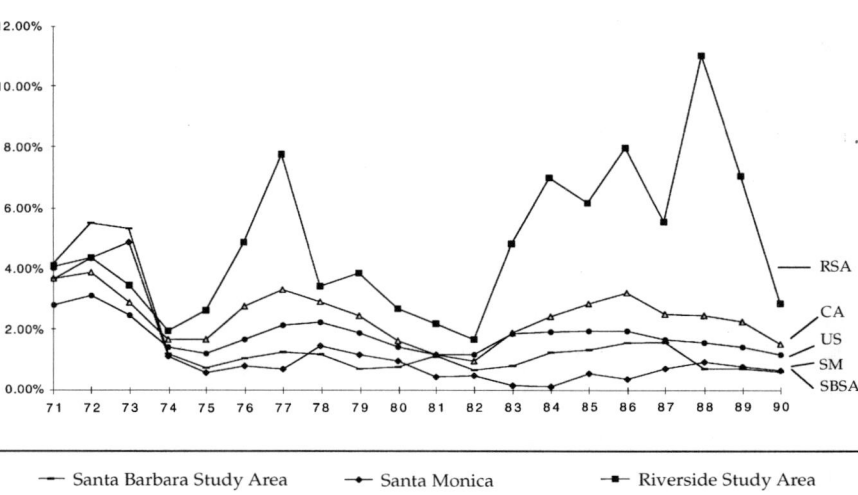

SOURCES: Riverside County, 1970–1989; Santa Barbara County, 1970–1989; and
U.S. Department of Commerce, 1970–1989.

The major drops in all areas came with the 1974 recession, brief in the
Riverside residential building case, but more persistent in the other two
areas (and in the state overall). The mid-1970s was also a period of tightening controls in Santa Barbara, suggesting that growth regulation
slowed residential building. But in Santa Monica this major drop in residential construction occurred before growth control (or rent control) became an element of local policy. The overall pattern is thus ambiguous, at
least if we use the kind of crude scan that Figure 3.2 allows. We are thus
left with the overall observation that state and national trends are important factors but that local growth controls might have had some effects.
Below, we discuss more refined techniques we used to measure precisely
the impacts of growth control as compared to the general trends.

Although growth was robust both in terms of commercial and residential growth in all our study areas, we notice that residential building rates
are much higher in the Riverside area, though the rate of commercial development in this area (whether in terms of retail sales or commercial
building) does not stand out from the other study sites. Compared to the
other study areas, Riverside's development has been skewed sharply toward residential construction, despite local efforts to bolster the city's

and county's fiscal and employment base, and despite the residential focus of most growth control in this area.

Testing the Power of Growth Restrictions

The growth in retail sales, commercial construction, and residential development does not demonstrate that growth controls have no bearing on rates of development. Such a demonstration requires more precise methods. Past studies, trying to pinpoint growth control impacts on such characteristics as population growth, rates of construction, and the cost of housing (or some combination of these) yield mixed results. Early studies tended to show that there were indeed impacts, findings quite opposite to the more recent analyses that fail to detect much by way of consequence.[2]

Reasons for the inconsistencies are the varying methods and different levels of precision used by researchers. An old standby method is to focus on a single locality, or a group of localities, and observe that growth declined "after" new controls were enacted.[3] Even when presented in a somewhat refined manner, these longitudinal studies run the obvious risk of attributing changes to growth control that may have occurred across various localities—including those without controls. Economies do go down as well as up, and all ships may decline together on the common sea.

Cross-sectional studies, which compare places with and without growth control at a single time point, may attribute effects to growth control that are due to some other unmeasured characteristic shared by places that have adopted growth controls—for example, their regional geographic location.[4] Common declines (or upswings) may be identified with growth controls when they are, in fact, a regional trend. According to Schwartz, Zorn, and Hansen (1986; also Schwartz and Zorn, 1988), the optimal study design is to compare across localities and over time, by way of a cross-sectional time series—an approach gaining increasing use, including by us (see, e.g., Appelbaum, 1985; Rey, 1988).

A second methodological problem has been the way researchers have identified "growth control" as a variable. The crudest approach has been to simply lump together all places that have some new way of regulating growth or that have the words "growth control" written into some legal measure or stated as part of local policy by a staff person answering a questionnaire (Elliot, 1981; Katz and Rosen, 1980, 1987). This approach blurs great differences in the content of various local polices, not to mention how well policies are carried through in daily administrative practice. A more discriminating approach is to at least distinguish between different types of growth controls. For example, Logan and Zhou's (1989) national study considered five distinct sorts of growth control. Still, it is the precise content of policies and implementation procedures that determine whether or not places "really" have growth controls. For example, some of Logan and Zhou's growth control policies (such as requiring environmental impact reports) are mandatory in some states but are locally enforced in ways that vary widely. Some localities merely "go along" with formal procedural requirements (and thus have all the right rules on their books), whereas others comply enthusiastically with the intent of state regulation. The distinction means a great deal both for developers and environmentalists.

Other researchers have scaled the degree of growth control based on survey responses that report the number of restrictive policies that have been adopted (Baldassare and Protash, 1982; Glickfeld and Levine, 1990; Neiman, 1990). Such "counts" of controls do not solve the problem of implementation differences. Passing numerous measures does not ensure their intensity or effectiveness. City councils can enact lists of measures with a single vote, simply incorporating items borrowed from another area's general plan or from a consultant's off-the-shelf inventory. In these cases, the proliferation of new policies may be more symbolic than substantive. For example, boomtown Corona would seem to have a higher level of growth control than the city of Santa Barbara, judging strictly from a tally of growth restrictions based on city officials' responses to a California League of Cities survey (reported in Glickfeld and Levine, 1990). This finding contradicts not only our field observations but the judgment of virtually any informed observer from either side of the growth debate.

An Alternative Testing Strategy

Whereas other studies supplement quantitative analysis with more descriptive case material (see Logan and Zhou, 1990; Neiman, 1990), we reversed priorities and used our knowledge of the local contexts to pick the policies we would take as the most important growth measures. Rather than treating growth control as a policy that can be read off from a questionnaire or county plan—as a uniform or clearly scaled phenomenon—our strategy was to use the field interviews (combined with plans and other legal documents) to find our independent variable. By interviewing a diverse set of local informants we were able to use agreement from political opposites as a signal of meaningful growth controls. Although the fifteen measures we identified seemed to vary in comprehensiveness, technique, and rigor, we made the assumption that those most involved in the local scene would have the best sense of which regulations mattered.

Having generated our "independent variable" in this way, we entered these diverse regulations into an appropriate model (cross-sectional time series) to discern the local impacts of specific measures on residential building activity. We developed a technique to separate growth control impacts from the whole range of other local variations in development activity caused by social, economic, historical, geographic, and cultural differences. We also accounted statistically for statewide changes in building activity in order to separate out the affects of broad shifts in the business cycle or macroeconomic factors such as interest rate fluctuation (see Appendix A for details and complete results).[5]

Contrary to the idea that new development regulations had stifled residential building, we found that in all but two instances—both of which took place in the early 1970s in the Santa Barbara study area—growth controls did nothing to slow residential growth. Commercial development was also quite robust, though we did not perform comparable statistical testing. It appears from our combined findings (and those of others)—and this is a conservative way to put it—that far more commercial growth occurred under growth control than local controversies suggested.

Our next task is to try to learn why so much development took place in the face of so many regulations presumed to curtail it—a topic to which we now turn.

Notes

1. Expressed in terms of the value of permitted construction. Data collection was facilitated by the government publications staff of the UCSB library and through the cooperation of the Construction Industry Research Board.

2. For early studies, see Dowall and Landis, 1982; Katz and Rosen, 1980; Urban Land Institute and Gruen, Gruen, and Associates, 1977; see Fischel (1990) for a review of the literature on price effects.

3. See, for example, Frech and Lafferty, 1984; Urban Land Institute and Gruen, Gruen, and Associates, 1977; Mercer and Morgan, 1982.

4. For examples of cross-sectional studies, see Dowall and Landis, 1982; Elliot, 1981; Katz and Rosen 1980, 1987; Wolch and Gabriel, 1981).

5. We used dummy variables to capture these broader sources of variation, as described in Appendix A.

4

Power to Build:
How Cities Grow
Under Growth Control

Growth control is apparently not all its cracked up to be, either by critics or supporters; formal policies have not pulled up the drawbridge. A generation of political science studies have taught that successful implementation of any sort of public policy can not be taken for granted.[1] Since land development is such a highly contested and complex operation with well-organized sponsors, regulatory outcomes are especially uncertain. Drawing upon various literatures on power, policy, and organizations, we offer four explanations for why prescribed growth limits might not take hold. We then provide more specific examples from our case studies, indicating how land-use regimes can function, using one broad mechanism or another.

1. Symbolic Politics. Growth control may be another arena in which policy makers, under constituents' demand to "do something," create formal rules and a lot of talk about regulation, but without real enough teeth to have material consequence (Edelman, 1964; see also Rubin, 1989).
2. Episodic Intervention. A policy may be applied only sporadically; with land development particularly, new policies must be consistently applied to produce intended results. The physical construction that may happen during "open" moments can hardly be repealed to restore prior landscapes. If developers slip through periodic windows of opportunity, controls only shift the distribution of approvals over time but do not reduce overall growth (see Janzcyk and Constance, 1980). Some rules are designed to be short term, often in response to some immediate crisis or shortage. Others turn out to be temporary because of shifts in political control or administrative practice.[2]

3. Countervailing Policies. Complex organizations, whether pri-
    vate or government units, have contradictory internal tenden-
    cies; some theorists conceive organizations as anarchies, with a
    "garbage can" of goals and strategies (March and Olsen 1976:
    175; Scott 1981: 270–275). Cities may be restraining develop-
    ment with some policies (picked up as growth controls by ana-
    lysts, developers, and environmentalists) and simultaneously
    encouraging it with others (not on the analysts' lists, not com-
    plained about by developers, although possibly a grievance
    among environmentalists).
4. Initiatives of the Regulated. We know from past cases of regula-
    tion (Huntington 1952; Lowi, 1969; McConnell, 1966) that the
    regulated have been able to function well and have even suc-
    ceeded in turning their regulators into advocates for their indus-
    try (e.g., farm programs, the TVA, and meat inspection). As in
    these other realms, regulated developers are not passive in the
    face of local growth control. Developers may take advantage of
    unforeseen slack, loopholes, or opportunities in the rules in order
    to get around them. Certainly, creative resistance and agency,
    which scholars now recognize among marginalized groups, are
    well within the repertoire of those accustomed to power.

We now draw on our case materials to show how these possibilities
work out in practice.

Growth Through Symbolic Growth Control

Political leaders often make general statements favoring some form of either "control" or "management" of growth, even when their policies
favor virtually all development initiatives. "Taking action" by adding a
"growth control element" to a locality's general plan opens a whole
episode of study reports, public hearings, and discussion. But, in the
United States, general plans are not binding in the absence of implementation measures, such as zoning (Popper, 1981; Rose, 1979: 53). In the
Santa Barbara County case (but not the city's), growth management implementation lagged eleven years behind the adoption of growth management planning goals. The county's "plan" for growth management
gave the impression of a control regime, but without implementation
procedures it was largely symbolic.

Plans often list community growth management "visions" as a substitute for strictures. A Santa Barbara County planner noted that translating
general "shoulds" to specific "shalls" was a large part of a revision of that
county's general plan.[3] But switching verb forms is not always enough.

For example, rather than limiting growth or growth rates, the Riverside County Growth Management Element did little except use a tough verb: "The County Administrator *shall* inventory and quantify current and anticipated social and economic needs in the County for 20 and 5 year periods. The Board of Supervisors, with appropriate public input *shall* be oriented to meet those needs" (Riverside County Planning Department 1990: 27—emphasis added).

Even when specific planning standards are set, they may still be inconsequential. The Santa Barbara County general plan update of 1980 was concrete enough to define specific thresholds of adverse traffic impacts that would be used to deny projects. But the county supervisors approved offending projects anyway, citing "overriding considerations" of jobs or housing—either one or the other is available in virtually all localities, given chronic shortages. Approving more jobs-producing projects then leads to a housing shortage; getting ahead on housing would (and does) lead to claims that more job-rich projects are needed to provide local employment to residents (the need for "jobs-housing balance").

Even the establishment of what appear to be "hard" population limits can turn out to be essentially symbolic. The city of Corona (in Riverside County) set a yearly limit of 450 housing units, a figure that exceeded the annual average over the previous five years of rapid growth. Corona exempted annexed land from its increased lot-size requirements, providing still another growth avenue. The city of Riverside's sewer limits also excluded annexed land, thus allowing an annual average of 2,309 new units to be approved instead of the 1,200–1,500 hookups per year under the rules.

Santa Monica lowered its zoning (in 1988) to permit no more than 70 million square feet of new commercial development, but that was still the equivalent of about fifty regional shopping malls and an amount ten-fold its previous 1984 build-out planning projection. Almost 6 million square feet had been approved by 1989, far surpassing prior expectations, but still not coming near the development cap. The boom led to later moratoria, which still exempted many projects. In Santa Barbara's case, the commercial build-out cap was low but did not restrain build-out in the near term. During this early phase, growth control was not really controlling; if later city councils were to relax the holding capacity to let development have another run, no real control would occur over the longer term.

Some of the new rules that appear to hinder growth in fact pave its way. For example, there can be measures that allow development only when fees are established to offset specific project costs. Such policies, as well as the fees that follow, may be loudly condemned as interference in free markets. But land development always presupposes government intervention (for infrastructure and basic services); when tax revenues are inadequate for the purpose, special fees and exactions facilitate rather than hinder growth (see Bauman and Ethier, 1987; Nelson, 1986; Snyder and Stegman, 1988). Responding to environmental groups' complaints of "unregulated growth," developers may also welcome such growth-enabling regulation, especially when the market is strong and when standardized fees allow relatively simple development review (Deakin, 1988; Alterman, 1988). As a prominent Riverside land-use and development lawyer said, "Developers may not like (impact fees), but they know they are rational. They figure in a boom, why argue?"[4] It is no surprise that these measures, sometimes castigated naively as "socialistic" or "confiscation," are unlikely to inhibit growth.

Growth Through Episodic Growth Control

Even a short lapse in growth control allows developers not only to satisfy pent-up demand but to build in anticipation of future regulation. Limits based on shortfalls in utility capacities are one type of control that tends to be temporary. Although control advocates may hope the infrastructure shortages will never be redressed, they often are.

Water policy in the Santa Barbara South Coast area seemed a deliberate use of infrastructure regulation to create long-term growth controls. Santa Barbara was the only county in the state whose voters had rejected the state water option (in 1979), in part because of the perceived negative environmental effects for northern California, in part for fear of how new water sources would stimulate local growth. Local growth interests, including a forceful local daily newspaper, had backed the measure strongly. But in 1991, after five years of drought, with brown lawns and short showers, citizens in almost every local water district voted to join the state water project (and, in a number of cases, to participate in desalination). The area's water districts developed vast amounts of contracted supply, eliminating any environmentalist litigation against projects based on their potential for draining this local resource. Thus, one of the few growth control measures we found to have had a measurable impact on housing production (the water moratorium in the Santa Barbara suburb of Goleta) turned out to be temporary, albeit twenty years in duration. Even in a strong version and in an environmentalist locale, resource-based growth constraints do not necessarily endure.

In the more pro-growth Riverside area, the city of Corona abandoned residential construction limits when it became possible to hook into a regional sewer line. In part because of the absence of infrastructure constraints, the same city dropped its citrus grove protection program eleven years after enactment, allowing developers to construct 12,500 houses on the designated land, a 66 percent boost in the city's total housing stock.

Another way growth continues after it has been "stopped" is through the "vesting" of rights to build projects that are "in process." Liberal interpretation of vesting extends weak regulation into the periods covered by tighter controls. With a 1988 Santa Monica commercial downzoning, developers were given ten days to complete applications under the old rules, creating a standing inventory of vested development rights. Goleta's growth management ordinance and the city of Santa Barbara's 1989 commercial growth ordinance limited vesting to those commercial projects already under review (but not yet approved).[5] Developers anticipate changes in regulation and try to "get vested" before a new regime takes over. Local lore had it that Santa Monica contractors were pouring foundations by lamplight (to meet the local vesting criterion) on the eve of the 1981 election that would bring an environmentalist slate into power.[6]

Another way of extending weak controls forward into future regimes is through development agreements—a mechanism created by the California legislature in 1979 that allowed localities to specify conditions for particular projects, specifications that could remain in force for up to thirty years, during which time the project could be built, remain on hold, or sold as a "currency" to other entrepreneurs who would buy the agreement when they bought the vacant land (Porter and Marsh, 1989; Pincetl, forthcoming). Once entered into, the agreements would bind the jurisdiction to the developer's plans (and the developer to the jurisdiction's conditions) on matters such as allowable densities, fee structures, and exactions. Projects were often set up to be built in stages over long periods (a decade or more; state law allows as much as thirty years). The absence of the kind of expiration date that ordinarily accompanies a planning approval shelters developers from changing political winds (and evolving ecological knowledge of their projects' impacts). Builders in Riverside County rushed to get under the wire in the months before the countywide growth control initiative (defeated, as it turned out) and secured development agreements for 100,000 future housing units.[7]

Under this same mechanism, Santa Monica approved an upscale 1.3-million-square-foot staged office development because of the new exactions the city was able to negotiate (opposed by some activists as part of a continuing "devils bargain" with developers). Another case was a pro-growth Santa Barbara County Board of Supervisors' approval of a "time-release" expansion plan for a Goleta research and development park that future slow-growth boards would be powerless to curtail.

The magnitude of development "in the pipeline" through one mechanism or another was often vast, making the effects of new growth controls a distant prospect. The 100,000 units approved by developer agreements in Riverside County were only a portion of some 150,000 housing units that, in 1989, were pending in the cities and unincorporated portions of western Riverside County—a number larger than the 102,000 houses existing in all the cities of this study site in 1980. Likewise, the 2.5 million square feet of retail space approved or under way in the city of Corona would have doubled the existing stock.[8]

At the level of individual projects, proposals for a given parcel or project were typically submitted repeatedly in different variations until the circumstances, including the political ones, allowed approval. A proposed seaside hotel development west of Santa Barbara, beyond the county's designated urban limit line, was twice denied under two separate ownerships but approved after the Hyatt Hotel Corporation entered the picture, beefing up mitigations and investing heavily in an unparalleled public relations blitz. In the case of another repeatedly rejected Santa Barbara area beachfront project, the Hyatt Corporation was unsuccessful in its efforts to win approval (this time for a project clearly within the urbanized area); instead, local developer-actor Fess Parker (of Davey Crockett fame) eventually mounted the successful campaign. Still another example from Santa Barbara was voters' rejection of a redevelopment subsidy to construct a downtown department store. The city council subsequently moved the project four blocks south and more than doubled both the project's size and the amount of the city's subsidy. They claimed the new site, which was redeveloped to include two department stores and connecting shops, better accomplished the city's "revitalization" purposes. The new location spared the local bookstore, whose owners had put up the majority of funding (about $26,000 out of their non-deep pockets) to defeat the first scenario.

Growth Through Countervailing Policies

Localities often limit growth in one realm but promote it in another. At the level of discourse, it is a difference between promoting growth (bad) and supporting "economic vitality," "redevelopment," or "revitalization" (all good). Developers can gain special treatment by connecting their activities to the good-sounding phrases and keeping distance from words that would imply they are doing something that might be "growth inducing." Santa Barbara pursued downtown "revitalization" using tax-increment financing (a $31 million public subsidy), despite a study's findings that the city already had California's second most prosperous retailing center—a fact that the study's UCSB professor authors (an economist and a sociologist) could barely get into public discussion (Appelbaum and Shapiro, 1983). Using the same redevelopment funding source, the city itself paid, on behalf of the department stores, the hefty traffic fees that it ordinarily charges for downtown projects ($2,000 for every peak-hour trip generated). The city was promoting development by giving back with one hand what was exacted by the other.

Playing a similar game, Santa Monica expedited project review for developments in its Third Street Redevelopment District (the "Promenade"), investing in streetscape improvements (paving, lamps, sculpture) and promoting new business with entrepreneurial zeal—all with great success, as it was to turn out.[9] The result was a "flourishing" downtown center that was to become "a regional destination point and a national model for downtown revitalization" (Rainey and Moore, 1996: B6). One reason for the high-level economic achievement of the city's "anti-business" city council is perhaps that its left-intelligentsia leaders' tastes led them to promote the kinds of facilities (theaters, restaurants, markets, boutiques, galleries) that would transform the city into an up-scale shopping and entertainment zone. It would not be the first time that left-wing intelligentsia set consumption patterns for those of different ideological leanings (the Bauhaus school of modernism was a prominent case). Far more than the traditional business leadership, this putatively anti-business group had the kind of cultural capital that would, ironically, yield the richest economic outcome. This city of left activists saw its credit ranking rise, becoming one of only two California cities to earn a Triple-A bond rating (Rainey and Moore, 1996: B6).

A political asset of using the redevelopment mechanism in any California city, but especially anti-growth cities, is its relative invisibility. Technically, decisions are made by a special authority (the "redevelopment agency"—often the city council wearing a different hat for the meetings). Its funds come from an earmarked portion of the property tax that is set up to be out of reach for any other type of expenditure. So it appears to the public to be spending found money; local media simply report that a given project or subsidy is being accomplished "with redevelopment funds" or "by the redevelopment agency." The process obscures true costs and insulates against alternative spending demands (see Friedland, Piven, and Alford, 1978).

Santa Barbara's severe tightening on commercial building was accompanied by a program to allow a transfer of development rights (TDR) from parcels outside the downtown core to the city center. Although this innovation did not formally increase the total amount of square footage that could be built, it could have that effect, because it permits owners of less desirable commercial locations to transfer their allowed densities to prime sites. Marginal commercial structures that otherwise might be cleared or converted to other uses (e.g., recreation, religious, educational) might be "reborn" in another locale and expanded to the limits of the law, thus increasing aggregate commercial development. Even as we write, we can report that the commercial cap policies are being further liberalized through other tinkerings.

Local governments promote growth through their tourist and convention promotions, partly using hotel bed taxes—the toll that visitors pay as a sort of special sales tax on motel/hotel charges (ordinarily between 7 and 10 percent). Again, by earmarking revenues for these purposes and turning the money over to a private organization (e.g., the local Chamber of Commerce), growth subsidies are insulated. At a more subtle level, governments promote local business through festivals, parades, and film company shoots; these activities and expenditures are rarely considered part of any local growth policy.

Growth Through Developer Initiatives

When growth interests do run up against rules that block their path, they often litigate. A group of landowners denied water hookups under the Goleta moratorium sued the slow-growth water district; with the help of a newly elected pro-growth majority on the district board and after sixteen years of court wrangling, the litigants gained a rich settlement. The district was obligated to provide water (even under the moratorium) or buy their lands at the with-water market value, an amount that would have bankrupted the district (Dalton, 1989: A1).

Litigation and the prospect of litigation can indeed work to slow down a slow-growth regime. A carpenters' union suit against Santa Monica's arts/social services fee (ultimately dropped) caused the city to negotiate its exactions project-by-project, rather than through blanket fees. Similarly, instead of risking court challenge, Santa Monica officials negotiated an office mitigation fee for parks and housing that developers "could live with."[10] Legal worries can also delay implementation for years as localities mind the procedural details, in the meantime providing for various exceptions to policies on the books. Proposed zoning changes in Santa Monica were postponed for seven years while the city's growth management plan was being reworked to make it more "bulletproof."

In still another legal strategy, growth interests used the very tools designed to restrain development in order to help their projects. Landowners in the city of Riverside challenged citrus-belt protections on the grounds that the city's general plan was inadequate. They cited such problems as its lack of density standards for commercial development—a claim that contradicts their usual resistance to such regulation. Likewise, the local Realtor association sued to block the 1989 Santa Barbara commercial cap on the grounds that the city had not conducted adequate environmental review. Litigation drains the public budget and makes local politicians vulnerable to the claim they are "bankrupting" the city with their fights against property rights.

Developers argued frequently that new regulations made projects unfeasible and unfairly deprived the developers of their assets. They relied on the legal principle that land-use laws must permit some viable

economic use of land or they are an unconstitutional "taking" of property. Decision makers often responded sympathetically to these claims
and wanted to be "reasonable" in the face of developers' hardship pleas.
Since local governments rarely had independent analyses to evaluate
whether a project really needed to be as large as developers claimed, they
relied on developers' calculations.[11]

Hardship claims related not just to project size but also to the allowed
uses of land. Some claimed that current uses were no longer economically viable. Landowners in the Riverside area complained that cities
were creating "brown belts" not green belts with citrus downzoning because the orchards were not economically viable. The same reasoning
was used in Santa Barbara County to argue that flower farmers should be
allowed to sell out to developers, or at least bolster their profits ("save
agriculture") by subdividing a portion of their land. In these cases, feasibility analyses are usually based on an inflated estimate of land value
that presumes a relaxation of planning restrictions. Policy makers often
then conclude that lower density development is not feasible because the
land values are too high to support lower-scaled projects. But values are
inflated because of investors' perceptions that development rules will be
relaxed. Weak enforcement in the past "sends a message" to the market,
driving up costs of land that would otherwise be cheaper. The policy
makers themselves create the value of the land, which they then feel
obliged to treat as a "condition of the marketplace" to which they must
respond with a policy exemption for the developer. The more they do it,
the more they feel a need to continue doing it.

A choreography of size negotiation is common as projects move
through different stages of gaining approval, at least in the tougher jurisdictions. Almost as a routine, developers initially propose projects they
presume will be too large to gain approval, even testifying that anything
less would be economically unfeasible. But when "starting with big numbers" is only a bargaining strategy, the formal "scaling down" that follows is not necessarily evidence of government-imposed limits, since the
higher density project was something of a fiction in the first place. Even
under the relatively tight restrictions of Santa Barbara, some of the projects that gain approval never do get built. This is another indication that
supply has not been choked off: The market cannot support what is already approved.

Developers often use the issue of economic feasibility to leverage approval of those elements most objectionable to decision makers. In
Moreno Valley where the purported "jobs-housing imbalance" means
more jobs are needed to balance "too much" housing, the developer of a
mixed-use project said he needed to develop housing, and to provide it
first, so there would be customers for the commercial/industrial portion

of the project that the city did in fact desire.[12] Alternatively, when cities want to encourage housing (for example, to achieve a "twenty-four-hour downtown," as was Santa Monica's goal), developers argue they must be allowed more commercial construction to support the less profitable housing.

The doctrine of jobs-housing balance thus operates as an all-purpose palliative. Business expansion remedies long commutes in the Riverside area; those sorts of projects not only gain easy approval but can leverage the housing construction that the city does not want but the developer does. In Santa Barbara and Santa Monica, housing moves most easily through the process (even with higher housing densities than are ordinarily allowed) because it means homes for local workers. And, once again, a developer can sometimes use the housing, especially if "affordable" (see Chapter 6), to induce permits for other types of construction as part of a mixed-use deal.

In all areas, developers sell projects on the negative argument that, whatever their disadvantages, if they are not approved they will be "lost" to a different jurisdiction; the Riverside area has been an especially fierce site of such competitions. Sometimes the claim is made, perhaps appropriately, that if the project goes next door the disadvantages of pollution and traffic will be felt, but the tax dollars will be lost.

Beyond generic claims, creative entrepreneurs find unforeseen opportunities to build within the envelope of new regulations. Growth grows in the legal and institutional cracks. To get around the Goleta water moratorium, developers argued, with some success, that the private wells they could drill would not tap public ground water basins. Then when a pro-growth majority won temporary control of the water board, the owners were allowed to turn their private wells over to the district in return for permanent water meters—a two-step process that bypassed the moratorium. Developers adapted to the city of Santa Barbara's ban on increased water use by using new water-saving techniques like indigenous landscape materials, low-flow shower heads, and toilets that flush with less water. They built projects using less water than the uses they replaced. In this case, a rule that otherwise could have stopped development instead allowed the city to conserve a resource.

A less constructive form of developer activism is simply to start projects illegally. Some Riverside County developers have allegedly plowed their lands under to eliminate discovery of an endangered species or of archaeological remains.[13] When the Santa Barbara Fire Department ordered developer Fess Parker to clear fifteen feet of weeds along the boundary of his land, he took the opportunity to bulldoze another five acres of wetlands and to remove illegally two acres of trees on his protected ten-acre parcel (Weston, 1990: B1).

Another pattern is to agree to development conditions but not fulfill them. Since public agencies can inspect only gross violations, end-users can easily avoid honoring project conditions (e.g., employers may not follow through with the ride-sharing programs promised by the developer who sold them their building). A Santa Barbara study revealed that although planners consistently underestimated negative impacts of approved projects, the conditions they imposed were often ignored in practice (Community Environmental Council, 1987).

One telling incident occurred in relation to an auto dealership development in Goleta. The project faced denial on the grounds of traffic problems and other aspects of over-urbanization at its site. To win approval, the property developer pledged to adjust hours of operation to mitigate traffic problems and agreed to pay the county to monitor compliance with the condition. But the car dealer who eventually moved into the structure (a different party than the developer) complained bitterly that the county's hyper-regulations were costing him business. The local newspaper ran a photograph of a county employee observing the building while monitoring compliance, generating negative editorial comment and angry letters to the editor about taxpayers' dollars paying for a county employee to "stand around." Lost in the controversy was the fact that the county created the arrangement to facilitate a project that otherwise could not have been built and that the hapless employee's time was not being paid for by the public. The monitoring was dropped by a county planning department in hasty retreat, rendering moot a series of imposed restrictions.

This is an example of how enforcing development conditions—especially complex and intrusive ones—can become difficult, especially in the face of an unsympathetic press and business community. Such conditions, even if potentially expensive, will not inhibit growth if developers suspect that they (or their tenants) will never pay for them. Even in Santa Barbara, many projects were insufficiently mitigated, either because impacts were underestimated at the time of approval or because inadequate monitoring had taken place (Community Environmental Council, 1987). This is but a symptom of the larger truth that implementation depends on consistent decisions of the "line" departments (e.g., public works, waste water management, public safety) as well as sustained political support and adequate monitoring budgets, a combination difficult to come by even in environmentalist jurisdictions.

The Context of Control: Comparative Power Advantages

The systemic nature of growth advocates' interest in land-use outcomes guarantees that under virtually any local circumstances they will remain involved on a continuous basis (see Alford and Friedland, 1985; Stone, 1976, 1981). Here, we discuss the general features of growth-coalition power that operate under the adverse circumstances of growth control. The larger development companies have staff committed to achieving political goals. A successful development team now may include planners, environmental analysts, land-use attorneys, public relations experts, various engineers (i.e., soils, structural, and traffic), and sometimes other specialized consultants (hotel experts, etc.). Environmentalists may have been the source of the original regulations (although see Walker and Heiman, 1981), but developers have effectively mobilized more than matching expertise. Smaller development firms and their consultants are active in support organizations such as the building industry associations, chambers of commerce, local development authorities, boards of Realtors, and ad hoc associations. In part through these groups, these firms gain appointments to various commissions and task forces that also help fashion local attitudes and policies.

The representative of a large development firm in Moreno Valley described his full-time job as "keeping touch with the pulse of the community" and claimed to be the only person other than city employees to attend all of the meetings of the city council and planning commission. One of his slow-growth opponents said, "He's paid to be there. He sits in the front row every week and is up there pinning the corsages on the women council members when they win an award so that they won't be able to look him in the face and say no to his project." A project planner for a development company described his job by saying, "I go to all of the people that say no and get them to say yes."

Such developer participation pays off. When a $7,100-per-unit impact fee was proposed for Moreno Valley, the building industry association responded with a white paper by its own expert. The 4,000 housing units that were approved while the issue was being debated paid only $1,200 per unit in impact fees—$23.6 million below true public costs as estimated by the city. Eventually, the fees were raised to $3,100, but meanwhile, developers saved $7.6 million.

Development proponents may know the city's land-use rules better than the city staff itself by virtue of their long involvement; sometimes, their own staff members had written the government's rules previously, when they were on the public payroll. We were told that planning directors hired from out of town or heading new planning departments (as in the case of several of the Riverside County cities) sometimes defer to developers' seasoned veterans, who can claim knowledge of policy intents and implementation traditions. Former city attorneys have become leading land-use lawyers in some cities (e.g., Santa Barbara); the former city manager of Santa Monica (chosen by the leftist majority) became a consultant for the most controversial development proposals in subsequent years.

On some occasions, growth opponents are able to hire their own experts; Santa Barbara's long-established Environmental Defense Center is the most prominent case in point. In an unusual Santa Monica case, a developer gave $2,500 so that a neighborhood group opposed to his project could hire its own traffic consultant (Fulton 1985, 105). Usually, citizen groups lean on whatever nonprofit groups may exist for technical and legal assistance, such as Legal Aid, the Sierra Club, the Trust for Public Lands, or on the planning organizations in place for this purpose. Women's groups, particularly the League of Women Voters and the American Association of University Women, play roles not only in their analyses of proposed policies, but also in acting as a proving ground for future public leaders. Sustained volunteer expertise varies among the areas and over time, but across all areas, environmentalists' participation is episodic compared to their opponents.

Sometimes developers can win out against prevailing policies by making a separate peace with particular subgroups that see special benefits from projects. These can be specific benefits like promises of aid for the homeless to placate social welfare advocates or a Santa Monica hotel developer's promise to fund beach patrols. Hyatt Regency gained support for its Goleta hotel from a surrounding tiny special district that would gain a tax bonanza. A Santa Monica office developer, using "consensual planning" (*Los Angeles Times*, April 22, 1990: K1) and offering various project amenities, won over the immediate neighbors, who then were less concerned with more remote difficulties (e.g., increased freeway system traffic).

More generally, growth advocates are finding ways to neutralize at least partially the "motherhood" appeal of environmentalism, constructing a political middle ground under the banner of prudent planning and growth management. The "moderate" developer position is usually marketed through grassroots organizations that developers create and sustain with names such as Citizens for a Balanced Community (Santa Barbara), Citizens for Reasonable Planning (Riverside County), Taxpayers for a Better Moreno Valley, and Santa Monicans for a Livable Environment (Clean Up Our Bay Beach and Parks Committee was another Santa Monica development organization). It is an open secret among all relevant players that the organizational impetus and most of the funding for these groups come from development interests. Despite claiming the middle ground, there were virtually no instances of any such ad hoc groups—or longer-standing developer association, local chamber of commerce, or daily newspaper—opposing any development project or measure.

In elections over land-use issues, growth forces usually outspend their opponents (Glickfeld, Graymer, and Morrison, 1987) on campaign literature, professional signature gatherers, market research, and campaign workers, both to oppose growth control initiatives and to elect sympathetic officials. Campaign experts advising developers recommend that they be prepared to outspend the opposition and to obscure contribution sources through PACs and misleading occupational designations. An article in *Builder* magazine (a developer trade journal) entitled "Winning at the Polls" laid these strategies out explicitly (German, 1990). Among the recommendations to counter citizen initiatives was one that called for mounting competing, but much weaker, measures on the same ballot, usually under the growth management, rather than growth control, rubric. These weaker measures (sometimes dubbed "the evil twin") are structured so that if passed by a larger majority, they take precedence (as was the case with a set of Riverside city ballot measures). Once voters are confused about which is which, an eleventh hour call may be made by pro-growth advocates to vote no on both, simply, as one informant said, "because the ballot is so confusing."

Some developers are careful to avoid an oppositional stance, tailoring their projects to local priorities and traditions—signs, perhaps, of a new modus vivendi. As a developer in the elite Santa Barbara suburb of Montecito observed, "Private property rights are no longer what they were. You cannot operate without sensitivity to the environment. Any landowner, developer, or builder who doesn't understand that is going to be left behind." (*Santa Barbara News Press*, February 10, 1990: B1). He took it in stride as the number of units in his latest project was cut by 20 percent and as forty acres were exacted as a nature preserve. Another Santa Barbara developer was at one time virulently against growth control, but later came to feel that "fights are never worth it." He decided to build his office and industrial complex around a historic structure rather than risk the wrath of local historians by demolishing it (*Santa Barbara News Press*, January 26, 1990: B1).

In Santa Monica, the entrepreneur behind a proposed beachfront hotel offered an on-site community center, public lockers, a ban on Styrofoam, and permanent funding for beach maintenance.[14] In a populist voice, he asked for a public vote: "The project and the people of Santa Monica will benefit when the narrow-minded, privileged special-interest groups which oppose this project are silenced by the people" (*Santa Monica Outlook*, May 5, 1990: A1). He and his consultants, former staff members of the progressive Campaign for Economic Democracy (Tom Hayden's organization), enlisted the support of school board members and social service providers. Badly split, Santa Monicans for Renters' Rights was unable to agree on a position.

When countywide growth control was defeated at the polls in Riverside, the heads of the Sierra Club and the local building industry association jointly requested some form of growth management program. The result was a plan that dropped the environmentalist-favored limits on the county growth rate and the general protection of agricultural lands but gained developers' support for development fees to fund a multi-habitat conservation area. Thus some environmentalists agreed to massive new growth in return for a modest set-aside fee.

Public officials and planners attend to developer preferences and are ready, indeed, to meet them at least halfway. The Santa Monica mayor, fearing his city's reputation might endanger financing of otherwise viable private projects, met with Los Angeles lenders to assure them their projects would be treated fairly. At the same time, an aggressive city manager recruited business leaders to help promote publicly oriented new growth (Clavel, 1986). This created a special opportunity for entrepreneurs not affiliated with the anti-rent-control interests to join policy-making committees and to participate in the more palatable development proposals, in some cases by tackling city-sponsored ventures.

In Santa Monica and Santa Barbara—and, to a degree, in the Riverside jurisdictions—the nature of growth control is one of negotiation and compromise, with even environmentalist officials keeping an eye on what growth interests will find reasonable, what they could litigate, and what they might seriously protest. This stance encourages adoption of mechanisms like caps that permit development to proceed without overt confrontation. Similarly, liberal interpretations of vesting spares those in authority the full ire of developers who can proceed with projects at hand. By making special allowances for projects that will employ people, officials can avoid the political fallout of employers threatening to move away if refused permission to expand. At a human level, there is a felt need to be responsive to those who, often graciously and reassuringly, make development requests day after day. Over time, the din of potential and experienced developer rationality, charm, litigation, and protest affects officials' policies and their implementation.

The Ways of Growth

Looking across cases, we find that the engine of growth, although distinctive in each locality, is a steady political and economic force; growth interests play a part at every point, creatively resisting and redefining restrictions. Where pro-growth sentiments are historically strong and oppositional organizations are weak, as in the Riverside area, growth controls leave room for development through their largely symbolic nature—broad proclamations and general goals. When specific standards are set, they again leave room for much development and, in some cases, give way under developer pressure. Utility access moratoria really are temporary, eliminated as soon as sought after supplies are gained. Local planners accept the need to compete with other communities near and

far to attract and subsidize business investment. Citizens resist the consequences of rapid growth but may settle for set-asides of conserved
open space. It is no wonder that—despite developers' complaints of
over-regulation—little happens in the way of aggregate limits.

This less ambitious regulation of land development is not easily explained by residents' class status. In the case of the city of Riverside,
household income virtually matches that of Santa Barbara (see Table
2.1). But in Santa Barbara, there is a citizen activist tradition that, as sometimes occurs in even modern communities, carries across generations as
"enduring difference" (Putnam, 1993: 136–137; see also Walton, 1992a).
Related perhaps to this difference in citizen activism are contrasting
levels of education; the proportion of residents with college degrees in
Santa Barbara (33 percent) and Santa Monica (43 percent) is much higher
than in Riverside (11 percent). These figures imply strongly that distinctions in political culture and organizational life between localities in our
study areas are more likely significant than family incomes in explaining
policy variations.

One of these distinctions is the capacity to organize and fund election
campaigns. Local regulations turn on election outcomes, both in terms of
referenda that enact controls and in terms of electing politicians who will
pass the regulations and see to their enforcement. We studied a series of
elections in all three of the central cities to determine where campaign
money came from and how it was distributed. The results, outlined in
more detail elsewhere (Lategola, 1992), show that in Santa Monica and
Santa Barbara the growth-control sides were able to raise serious money.
In Santa Monica they were still outspent by a better than two-to-one
ratio, but the approximate $100,000 they raised in the 1990 elections, for
example, permitted a sophisticated campaign. In Santa Barbara, the
growth-control side raised about the same amount, but in this case it was
a total that actually surpassed the fund raising capacities for their adversaries. Riverside environmentalists, in contrast, ran shoestring campaigns; even when they won, their opponents outspent them by ratios of
seven to one—not very different from the usual eight-to-one ratios in
such pro-growth versus slow-growth contests in California during this
era (Glickfeld, Graymer, and Morrison, 1987: 125).

Over the years of fights over land use (and rent control in Santa Monica), the total amount of money spent on local elections for council and
county supervisor seats rose sharply. For example, contributions rose 314
percent in Santa Barbara between city elections held in 1977–1978 and
those held in 1989–1990; for Santa Monica the rise was over 1,000 percent
(Lategola, 1992). These totals greatly exceed campaign cost inflation for
seats in the California State Assembly, which grew by a more modest 150
percent over the same period.[15] City of Riverside campaign contributions

grew not only at a lower rate than in the other two cities but also at only about half the rate (at 83 percent) of the cost increases for seats in the state assembly. The Riverside environmentalists, in particular, appeared unable to increase their fund raising capacity over time. Whereas their opponents increased funds by 147 percent (matching the statewide standard) between the two time periods, the slow-growthers' total campaign spending slightly declined. Unlike Santa Barbara and Santa Monica activists, those in Riverside were unable, for one reason or another, to mount increasingly expensive campaigns. Without money, a political group must fall back on some other extraordinary resource, like grassroots organizational strength, a historic tradition of loyalty to a particular party or cause, or perhaps strong backing from local media. None of these was present in Riverside.

One possible explanation that will *not* work to explain different election and policy outcomes among the three cities is the role of the local media, at least the daily press. In all three cities, the local daily paper strongly supported growth, both in terms of backing particular projects as well as in terms of opposing regulations that would curtail development, in general.[16] This sympathetic stance toward development was reflected in editorial positions; no growth control measure was supported, nor was any particular project opposed by any of the three papers over the course of the study period, as indicated both by our informants at the papers and by our own checking of newspaper microfilm on specific cases. Further, our own systematic examinations (Warner and Molotch, 1992: 101–102) of "straight news" coverage of particular projects showed a consistent tilt toward favoring the pro-development side (for example, by using pro-project sources in the news story but not environmentalists). Although there were some differences in media coverage between the localities in our study area—especially in regard to how much detail was published from environmental review documents (see Chapter 5), it is safe to conclude that when environmentalism did prevail in one of our cities, it did so in spite of the local paper, not because of it.

Although we are reticent to overgeneralize from case studies, it does appear that certain types of government actions are more effective than others in making growth control count. Symbolic commitments to growth control through general plan amendments or new wording in city charters clearly have weak impacts. On the other hand, controls based on concrete but intrusive and idiosyncratic monitoring, as in the auto dealership case, risk public disapproval and may damage the reputation of a whole regimen of environmental controls. Infrastructure constraints are also unlikely to be effective; expanded capacities can be developed, sometimes through demands from a public that, however initially opposed to development, comes to suffer when water (or road) capacity shortages hinder their life routines. Growth caps may also not be very effective, given the tendency (as is clear from experience in many areas) to "adjust" them when development starts bumping against imposed ceilings, or they may be circumvented in still other ways. Although the caps used in our case cities were not raised significantly during our study period, evidence from other places suggests that they likely will be. Most cities in California's Ventura County have officially capped the annual number of residential building permits, but these limits are consistently exceeded, with as many as fourteen times more permits being issued than allowed (Martin, 1991).

What, then, can work? Although we have less relevant data on this issue than for some of our other conclusions, it is likely that a combination of downzoning and a change in legislative voting requirements (a super-majority for upzoning) may have strong consequences. Such a combination helps institutionalize a fundamental condition for effective controls: a favorable political environment sustained over time. The presence of such political circumstances—much more than the number, type, or title of controls (indeed, whether the words "growth control" are used at all)—determines whether regulations matter in the way places grow.

Notes

1. Although the usual approach to the implementation problem is to look within bureaucracies for the dynamics that decouple legislative intent from policy outcome (see, e.g., Mazmanian and Sabatier, 1983; Pressman and Wildavsky, 1973), we include the surrounding "social politics" (Brodkin, 1990; Freudenburg and Gramling, 1994), involving those outside the official realm.

2. Besides the political transformations we observed at our study sites, Calavita (1992) reports that San Diego, a city once led into growth control by then mayor Pete Wilson, returned to growth machine dominance.

3. Interview 012. We refer to informants by interview numbers to assure confidentiality.

4. Interview 053.

5. It can also go the other way; certain of Santa Monica's 1988 commercial downzoning measures were applied to projects that had beaten the vesting deadline, generating threats of legal action.

6. Interview 026.

7. Although a source of pride for local planning officials, the agreed-upon $4,000 per unit impact fee built into these developments was relatively low; a Santa Barbara County planner-informant dismissed it as enough "to pay for the curbs" (Interview 004).

8. The reference year is 1991. See "City Focus, Inland Empire: Corona," *Inland Empire Business Journal*, January 1991: 54.

9. Interview 030.

10. Interview 029 (also see, Hamilton, Rabinovitz, and Szanton, Inc., 1982).

11. Santa Barbara County's Hyatt Hotel project is one instance where a study was required. See "Study Says 400 Rooms Needed for Hyatt profit," *Santa Barbara News Press*, July 7, 1988: A8.

12. Interview 007.

13. Interview 006.

14. The relative environmental costs of Styrofoam versus the alternatives have since come under much greater scrutiny (Hocking, 1991).

15. We used these time points for comparison because the California Fair Political Practices Commission did not provide data for earlier periods.

16. The local dailies are the *Santa Barbara News Press*, the *Santa Monica Outlook*, and the *Riverside Press Enterprise*. These were not the only media in the three areas, but on local issues they were the most important. Santa Barbara was also served by at least one weekly "alternative" paper over the era, primarily the *Santa Barbara News and Review*, which took a consistently pro-environment stance. Santa Monica was also served by the *Los Angeles Times*, but this paper gave less coverage to local Santa Monica affairs than did the *Outlook*. To a degree we have not determined, the Santa Monica-based public radio station, KCRW, played an important role in local controversies and was a liberal-left, pro-environmentalist outlet.

5

Project Peddling:
What Gets Approved and How

Controls do not, by any means, stop growth and in most cases are likely to have no effect on its aggregate amount, but this does not mean they simply have no consequence apart from encouraging grumbling by developers and some satisfaction for environmentalists. In this chapter and the next we assess how new regulations *have* altered development practices. In so doing, we will provide a better understanding, in quite precise terms, of the nature of the development process under tightened building rules. In the last chapter we saw the ways developers move around in this regulatory context. Now we want to understand the context itself. To do this, we identify what local governments gain from developers as they shape their projects to win governmental approval.

Clearly, we cannot rely on lists of formalized development requirements to understand how governments shape development (this we learned in previous chapters). Instead, we must take seriously and delve still more deeply into the implementation process—the way a particular project gets changed, modified, or subverted as it moves through to approval.

Our original goal was to create standard measures of the public benefits gained from development approvals that could be applied across localities, something like the dollar amount of public funds generated per square foot of construction—perhaps broken down by the type of development (e.g., office versus malls versus residential). Such indicators would allow us to compare the intensity and the content of regulation from one site to another. This type of calculation proved to be impossible. For one thing, local governments do not keep track of the total fees they collect per project or on a citywide basis. This is partly because some of the fees are imposed ad hoc in the form of mitigations required on specific projects (they are site specific, rather than coming from a uniform schedule of fees). There are also a wide array of agencies and special districts that collect fees and/or impose their own development requirements (e.g., school districts, sewer districts, building departments, park boards, redevelopment authorities, etc.).

More importantly, collecting fees is only one way localities offset the costs of growth. They may instead limit the scale of a project or redirect it to a location that reduces its major cost impacts. In effect, they may gain more in cases where they collect no fees. Further, developers themselves may be required to remedy adverse effects—for example, by providing mass transit passes to their workers to reduce traffic congestion (as opposed to the city collecting fees to widen roads for them to drive on). They may be asked to fund low-income housing that their project displaces or, as an alternative, to build affordable housing on site. Developers can be asked to pay money toward an open-space fund or can be barred from using grading practices that would harm wetlands. In fact, various mixes of mitigation and regulation may coexist within the same locality and be applied differently from year to year or from project to project.

Even the most mundane exactions—say, traffic impact fees charged on the same type of project—may differ in how they are assessed (per square foot, per housing unit, per average daily trip), where they are applied (in certain districts, for certain land uses), and how they are triggered (by certain size thresholds, by whether or not full environmental review takes place). Past surveys have helped show the range of development exactions and are useful for comparing places that are similar on specific dimensions (e.g., traffic or park land fees in two booming residential suburbs that use the same techniques of land use control).[1] But in most instances, comparisons between places are difficult without knowledge of actual project circumstances. If we simply count up the number of exactions or their recommended dollar value, we might conclude erroneously that the places with the greatest number of fees are also the toughest on development. This runs counter to the perceptions of local officials, such as a Santa Barbara planner who remarked, "We don't have impact fees, we are not into growth here."[2]

The different elements of a project also interact with one another: A project may benefit the community by providing low-income housing and also reduce the public burden of heavier traffic congestion by bringing workers closer to downtown jobs. So the project's net contribution to local welfare has to be judged in combination. To try to total local gains using lists of policy exactions hides the variety of factors at play and risks summing incommensurate and sometimes immeasurable items.

Peddling Around with Our Projects

To address this methodological problem we decided to compare how the same "actual" projects would be dealt with across localities. To accomplish this, we asked veteran city planners in the three main cities (Santa Barbara, Santa Monica, and Riverside) to review projects approved in the other two jurisdictions as though they were being proposed for their own locality. We generated the projects by asking each planner to choose two developments that had been approved in their city recently. One was to be "routine," representing a common form of development (in scale, location, and type of use) and with conditions of approval that were at the norm in terms of type and level. The routine case from Riverside was an apartment project; the cases from Santa Barbara and Santa Monica were office/retail mixes. We refer to these routine projects throughout this text by fictitious names to make them easier to recall—Barbara Plaza, Monica Tower, and Riverside Arms (see Photos on pages 86–88, 91, 93, 94).[3]

For the other selection, each planner nominated a "best-case" project where the locality secured maximum concessions from the developer.[4] Compared to the routine projects, these best cases were larger and mixed various combinations of office, retail, and residential development. To indicate the special quality, from the planners' viewpoints, of these projects, we use three stars within each of their designated names: Rancho***Barbara, Monica***Centre, Riverside***Place.[5] Brief descriptions of our project cases are contained in Table 5.1 and further detailed in Appendix E.

We then forwarded project descriptions and conditions of approval (the actual text that conveyed official action as well as our summaries of these actions) to the other two planning departments. We took the path of a developer looking for a building site, peddling the projects from one place to another. After each official had examined the materials (in one city with a full staff meeting) we returned for extended interviews with each planner. We asked them how each proposal would be received by their jurisdiction, whether it would be approved, in what form, and with what additional (or fewer) conditions attached. We also asked the planners to describe what might keep them from asking for more restrictions (political pressure, economic interpretation, technical preference); this information was often volunteered by our informants, a highly articulate and thoughtful group. We also asked about the conditions imposed on similar projects approved in their city (see Appendix F for our interview schedule).

What Gets Approved and How

The first thing we learned was which cities would approve which project as proposed. In Riverside, all of the proposed projects would have been

TABLE 5.1 Project Descriptions

Routine Projects	Best-Case Projects
Barbara Plaza (Santa Barbara) 801 Chapala—office/retail project one block from the main street, 6,000-square-foot, three-story building on a 15,000-square-foot lot. Monica Tower (Santa Monica), 1250 Fourth Street—office/retail complex downtown; 93,451-square-foot, six-story building on a 30,000-square-foot lot. Riverside Arms (Riverside), Spruce Apartments—102-unit, two-story apartment complex on 4.5 acres (196,020 square feet).	Rancho***Barbara (Santa Barbara)—two-story mixed-use project including commercial (auto dealerships, 32,653 square feet on 291,852 square feet of land) and residential (96 one- and two-story residential units on 9.65 acres—420,354 square feet). Monica***Centre (Santa Monica)—six-story, mixed-use project including 165,000 square feet of retail and 1,094,577 square feet of office on 17.1 acres of land (744,876 square feet). Riverside***Place (Riverside)—mixed-use project with a maximum of six stories, including commercial (500,000 square feet of office, 260,000 square feet of retail, 300,000 square feet of hotel/conference center on 55.3 acres, 2.4 million square feet) and residential (570 units on 42 acres—1,829,520 square feet).

approved for construction. Santa Monica would have accepted the Santa Barbara projects but denied those that originated in Riverside. Santa Barbara would not have accepted any of the projects from the other two cities as originally proposed. Thus, we found an ordinal scale of regulatory rigor. To some degree, this finding confirmed that the city of Riverside was weaker in its controls compared to the other two places. It also clarified the more ambiguous comparison of Santa Monica and Santa Barbara, each of which could, in its own way, vie for being the more stringent. The exercise provided us with a gross measure of how strict cities were relative to each other based on the binding public decisions that implement planning policies. Table 5.2 summarizes these results, distinguishing between the routine and best-case projects.

TABLE 5.2 Responses to Routine and Best-Case Projects Proposed to Each City

	Santa Barbara	Santa Monica	Riverside
Barbara Plaza	approve	approve	approve
Monica Tower	deny	approve	approve
Riverside Arms	deny[a]	deny[a]	approve
Rancho***Barbara	approve	approve[b]	approve
Monica***Centre	deny	approve	approve
Riverside***Place	deny	deny	approve

[a] A zone change from single- to multi-family residential would be unlikely.
[b] An exception would have to be made to the current commercial moratorium.

This comparison tells "what" is approved and denied but does not tell "why." To learn how and why decisions differed from one place to the next we questioned the senior planners, studied the official documents carefully, talked with other city staff, and reviewed local media coverage of relevant development decisions. First, we will compare the reactions to "routine" developments. Then we will turn to the best-case projects.

Routine Projects

Riverside Arms, Riverside's routine case, required a zone change from single-family to multi-family residential. This kind of "upzoning," which increases the level of development allowed on a site, used to be so common in U.S. cities that zoning was quite malleable (Babcock, 1969). Even though Riverside community groups often resisted such changes—whether they increased density or intensified the type of land use (e.g., from residential to commercial or industrial)—upzoning was common enough that this project could be regarded as "routine." It would have been denied on the basis of the zone change alone in the other two cities. Thus, one "new" challenge to developer power is for governments simply to abide by existing zoning and approved plans rather than to grant exceptions easily. Following the rules already in place is a powerful form of growth regulation that is not represented on lists of growth controls. Santa Monica officials came to consider single-family zones as a kind of "sacred land"—approving higher densities would be considered a "sacrilege" (see Photo on page 83).

Single-family areas in Santa Monica

Indeed, in Santa Monica, zoning reflected "the basic land-use pattern of the city" and was altered only by "a little tweak here and there ... for very unique reasons," certainly not as a routine matter.[6] Similarly, the Santa Barbara planner, speaking in 1989 of both zoning and the general plan, said, "Until the major general plan amendment last year we didn't process any general plan amendments since 1986, except to fine-tune some language in the different sections."[7] These comments indicate how rare it would have been to rezone land for any routine development project. Residential developments in Santa Monica and Santa Barbara either met existing zoning requirements or were part of a mixed-use project in a non-residential area. It is not surprising that the "routine" office and commercial projects, Barbara Plaza and Monica Tower, both conformed with existing zoning.

Santa Monica would have accepted Santa Barbara's routine office/retail project, and then some. Projects like the Santa Barbara project, with less than 15,000 square feet of floor area, were unaffected by any of Santa Monica's commercial growth controls, including the temporary moratorium of May 1989.

Indeed, a structure three times as large as Barbara Plaza could have been built on a comparable site in Santa Monica and still have been exempt from the moratorium. Monica Tower was built on a lot twice as big as that of Barbara Plaza, but the Santa Monica project had sixteen times more floor area, was more than twice as tall, virtually filled its site, and provided parking for 321 cars, compared to the 24 parking spaces at Barbara Plaza (see Photos on page 85). The dramatic density difference reflects the different scales of development allowed in these two cities, not a greater willingness to make exceptions to the rules.

Even before implementation of the city's 1989 commercial restrictions, Santa Barbara would not have approved the routine Santa Monica project. The city's planning commission exercised considerable discretion on all commercial projects, guided in part by the size of any preexisting structures (the more already present, the more that could be built to replace it) and by the history of water use for the site (with no net increases in water allowed). If these conditions were met, a developer in the post-1989 era might have been allowed an 11,000- to 12,000-square-foot building (about twice as big as Barbara Plaza, see Photo on page 86) on a 30,000-square-foot site, rather than the 93,000 square feet approved for Monica Tower (see Photo on page 87).

Riverside's routine residential project, which covered 4.5 acres, would have been an unlikely proposal in the other two cities because of the shortage of large developable parcels. But if spread across a number of different sites, its density would have resembled existing zoning in both places. Riverside Arms included 23.3 housing units per acre; Santa Barbara

Parking at Barbara Plaza and Monica Tower

Barbara Plaza

Monica Tower

Riverside Arms

permitted 20 units per acre and Santa Monica 30 (see Photo on page 88). But a developer would have to provide affordable housing in order to build at this density in Santa Barbara, and to build at all in Santa Monica (these affordability requirements will be detailed in the next chapter). Santa Barbara and Santa Monica thus permitted the same residential density (or more) than Riverside, but only if the cities' affordability stipulations were met.[8]

The amount of building allowed on a site can also be affected by other design regulations, such as building set-back requirements. Under Santa Barbara design review, multi-story buildings are usually stepped back from floor to floor so that third and fourth floors are only a third as large as lower floors, with four stories being the seldom-permitted maximum set in the city charter. In comparison, Santa Monica officials reduced the top floor of Monica Tower by less than 10 percent. Riverside would have made revisions of a similarly minor order on a project of this scale.

Best-Case Projects

Compared to the routine projects, the best-case developments hinged much more on case-by-case negotiations between city planners and private developers. This is where policy priorities were translated into specific development requirements. All of the three-star projects required land-use designations that were custom designed for their sites. When the city of Santa Barbara approved Rancho***Barbara for construction, planners maintained the same overall mix of land uses (commercial, residential, etc.) permitted on the site but modified the existing zoning to rearrange their distribution within the land parcel. The planner said, "the (zoning) line went one direction and we switched it to run the other direction."[9] In Santa Monica, the mixture of uses included in Monica***Centre also conformed with existing zoning but was confirmed and specified through a customized development agreement between the city and the developer; this agreement allowed the project to proceed. The project was in a special office district the city had recently created to upgrade a semi-industrial area of town.

In contrast with these two cases, Riverside***Place required more fundamental rezoning from the city; commercial and multi-family development was allowed in what had been a single-family residential zone. The Riverside City Council justified rezoning on the grounds that part of the development would address community educational needs. The Riverside general plan of 1969 designated the site as the new campus for Riverside Community College. Together with a private developer, the college, as the landowner, was proposing the new development. Actually, only 5 percent of the proposed construction would have provided public educational space; the other 95 percent was given over to other uses, including more than one million square feet of commercial development. Almost any rezoning from residential to commercial, much less of this magnitude, would be highly unlikely in the other cities.[10]

Santa Barbara officials thought they could accomplish a combination of goals by approving the housing and auto center that made up Rancho***Barbara (see Photo on page 91). They could retain sales tax revenues for the city (by preventing a threatened exodus of car dealerships from the city limits, and even adding more dealerships) and improve the traffic system at the developer's expense (by having developers finance a new freeway off-ramp and link two important traffic arterials). The city also gained needed affordable housing by granting "bonus density" so that the developer could build more housing units on the site (as a result, the housing density exceeded that of Riverside's routine project).[11] The site was an undeveloped parcel within the uptown urbanized area, buffered from nearby residential areas by the freeway, an existing shopping mall, and a retail strip. At this location, the project could address the three different problems simultaneously: keeping sales tax revenue, improving traffic circulation, and providing affordable housing. Although still controversial (in part because of possible noise from the dealerships and high housing densities), its precise configuration and siting rendered the project relatively inoffensive.

Development was permitted only after about five years of negotiation and the rejection of two earlier proposals. The developer wanted originally to build only offices, then proposed a hotel/convention center—something favored in the general plan (but downtown, not at this uptown site)—and finally came around to the auto center/housing concept. In effect, planning regulations that strictly limited the density of housing, especially if it was not "affordable," and a climate that made any zoning change difficult (even the internal rearrangement of zoning lines required for the project) pushed the developer to focus on the substance of the city's planning goals.

As with Santa Barbara, Santa Monica also approved its best-case project because of its fit with overall city planning strategies. The vast Monica***Centre commercial project would enhance public revenues and support the social services high on the city's agenda (see Photo on page 93). The project was in an area where planners welcomed development that would yield such revenues; few existing housing units would be displaced; no architecturally significant buildings were in its path; and relatively few nearby neighbors would be disturbed. Besides agreeing to a wide array of exactions, the owner consulted the city and neighbors regarding the kinds of amenities and uses to include in the project. The same approach was replicated and refined by later developers, with the advice of the former city manager (part of the earlier left regime), now acting as a private land-use consultant.[12]

Rancho***Barbara

Despite the fact that Santa Monica was becoming more wary of large-scale development, both the Santa Barbara and Riverside projects would have been considered attractive if appropriately zoned sites were available. Because its commercial densities and scale were relatively low for Santa Monica, Riverside***Place would have fit well into Santa Monica's city-owned airport site (see Chapter 6 for details on this project). The city would have welcomed the housing component if it included a defined portion of subsidized units. Santa Monica would have also encouraged Rancho***Barbara because of the affordable housing component and its auto dealerships (an acknowledged source of tax revenue). In Santa Monica, such a project probably would have been allowed at higher densities, since Santa Monica allows both larger scale residential and (especially) commercial development.

Riverside would have welcomed any of the best-case projects because of their commercial components. In fact, the projects could have been purely commercial—and housing, if included, would not have had to meet any affordability tests. Riverside's own best-case project met neighborhood opposition, but not because of the amount or intensity of commercial development. Instead, neighbors objected to the proposed apartments that would abut their single-family neighborhood. In fact, the city planning staff had recommended against such high residential densities but were ignored by the project sponsors who gained planning commission approval for twice the residential density recommended by staff. The neighborhood opposition spurred the city council to approve 40 percent fewer apartments. After our interviews were completed, the newly elected slow-growth mayor vetoed the project because of neighborhood concerns. The developer dropped the apartment component altogether to get final approval.

Although commercial densities tended to be lower in Riverside than in Santa Monica, this did not reflect a limit on commercial densities but a limit in the market for commercial space. Indeed, the Riverside***Place project was the one case study that was approved but not built because of weak market demand (see Photos on page 94). (It is interesting to note that the only project of our sample that was shelved came from a *less* regulated context.) Officials had allowed projects for the downtown area that were similar to Monica***Centre in intensity and scale. According to a Riverside planner, referring to commercial space in general, "No one is trying to scale the project down, on city council, in the neighborhood, in downtown."[13]

Riverside was even more eager to attract an auto center like Rancho***Barbara. Rather than restricting the size (or asking for affordable housing as part of the project), the city would have wanted the auto complex as big as possible. Auto malls in the broader Riverside region were

*Monica***Centre*

Riverside***Place undeveloped site

the objects of fierce competition between the various localities (Sunderland, 1989: 14–16) and were six to seven times larger than Rancho***Barbara, covering as much as seventy-five acres. To accommodate one proposed auto mall, Riverside City Council unanimously rezoned 150 acres of land to a general commercial zone. In the process the council threw out all the design and landscaping standards planners had previously crafted for the site. The lone voice in opposition was the neighboring city of Moreno Valley, which was trying to launch an auto mall of its own. This Riverside auto center was never built, but the owner retained the relaxed zoning for the site. As a result, the lost auto mall project became the vehicle for a permanent change in land-use designation; "they could build the world's largest open-air flea market," the Riverside planner lamented.[14]

In contrast to Riverside's welcoming of best-case projects from the other two areas, Santa Barbara probably would not even have approved Rancho***Barbara in later years because of its size and because of higher standards for project mitigations. Under Santa Barbara's 1989 clamp-down on commercial growth it would have been impossible. Riverside***Place would use up more than a fourth of the total commercial space available under Santa Barbara's growth cap through the year 2009—hardly a reason to grant it special treatment. Santa Monica might have been interested in Riverside***Place, but only with measures to build the local stock of affordable housing and mitigations of other sorts.

Project Environmental Reviews

One of the important aspects of regulation is the way localities carry out environmental review as mandated by the California Environmental Quality Act of 1971 (CEQA), the state's adaptation of the National Environmental Policy Act (NEPA) of the prior year.[15] For developers, such reviews determine how much money they will have to put up for studies, the costs that may come to them as a result of the studies' findings, and the degree to which study findings stir up publicity and public opposition to cause them trouble. The review process has the potential to bring authoritative evidence before the public both on the substance of the project and the adequacy of the environmental analysis itself. Through mandated hearings, it can provide opportunities for public testimony, the registration of diverse perspectives, and media attention.

According to state law, any project proposal that deviates from current rules—such as existing zoning or a development cap—has to go through environmental review. The same is true for any project requiring "discretionary" approval (i.e., anything beyond routine staff administrative approval, whether by board, commission, or council). CEQA also requires environmental review whenever a project provokes significant "controversy."

Once CEQA is triggered, local officials have three options according to California law:

1. Issue a "negative declaration," meaning that the city deems that the proposal has no significant environmental impacts and can thus proceed
2. Issue a "negative declaration" but with conditions, meaning that the project can go forward if the developer mitigates certain environmental impacts
3. Require a full environmental impact report (EIR) either to define the conditions for approval or, if mitigations are not feasible, to justify the disapproval of the project

These statewide criteria give localities leeway to determine the kinds of projects that would trigger CEQA review, as well as the kind of review that would take place under the law (i.e., which of the three review options would be taken). Local governments set the threshold of what can be approved administratively and, through zoning, establish the envelope of currently allowed construction. If the envelope is big enough, there would seldom be a need to "break" rules and hence to trigger CEQA review. There is also obvious room for local interpretation of what constitutes "controversy" in a given situation. The realms of discretion interact, in that places with large zoning envelopes and permissive general plans create less need for special action and also dampen controversy by creating the attitude that approvals are inevitable.

In our study areas, the delineation of what is administrative versus discretionary action differed across places in quite obvious ways. In Santa Barbara, for example, decisions by the Architectural Board of Review (required for a wide variety of projects) were defined as discretionary, triggering environmental review. Santa Monica and Riverside were less likely to require architectural review on a given project; they defined these reviews as administrative, *not* triggering environmental review under CEQA. We see that within the same regime of state law, local interpretation can be critical.

In Riverside, projects of whatever scale could be approved administratively as long as they conformed with current zoning. Given Riverside's large envelope of unused zoning rights, this meant that CEQA environmental review was invoked rarely. As late as 1980, Santa Monica had a similar policy and approved a 1.3-million-square-foot project "over-the-counter"—that is, without any public hearings, planning commission consideration, or environmental study. Such a project would later require

TABLE 5.3 Routine Use of Environmental Review

	Santa Barbara	*Santa Monica*	*Riverside*
Barbara Plaza	ND	none	none
Monica Tower	EIR	EIR	none
Riverside Arms	EIR	ND	ND

NOTE: ND=Negative Declaration, EIR=Environmental Impact Report.

traffic studies, architectural design review, and a series of public approvals. Santa Monica came to require environmental review on all of its larger projects (regardless of zoning conformance) but exempted many more from CEQA review than Santa Barbara. Santa Barbara had long used a strict interpretation of CEQA, exempting only residential projects of fewer than six units. Virtually all commercial or industrial projects required review for one reason or another.[16]

By introducing new limits and expanding the domain of discretionary action, growth controls bring environmental review to bear on more development projects. This tends to happen precisely in those places where environmentalists are prepared to engage seriously in the review process. Competent environmental review provides more complete information and better understanding of project impacts. This process then can kick off all sorts of media coverage, stimulating still more public controversy and demands for mitigations from developers. In all three cities, the best-case projects required environmental review because they involved some discretionary action (e.g., approving a specific plan amendment to the general plan, a zoning change, or a development agreement), but the typical use of environmental review clearly differed from place to place (see Table 5.3).

Riverside's requirement of environmental review for its routine Riverside Arms project came about only because a zone change was involved. Even so, they issued a negative declaration, specifying only minor modifications of the physical design. If the project had been proposed in an existing multi-family zone, no review would have been necessary at all. Likewise, the Barbara Plaza project and Monica Tower would have been approved in Riverside with no review. The Riverside planner could think of only one project within the city that had required a full EIR when there was not a zone change: the doubling of a regional mall (to 1.5 million square feet). As our informant told us, "If they have the zoning, all they're really doing is going for a building permit." Still the city of Riverside considers itself restrictive compared with neighboring jurisdictions. Our Riverside planning informant commented, "Most of the places around here say, 'Environmental review? Yeah, I think we have a stamp around here for that.'"

Santa Monica required a full EIR for its own routine project but did not demand mitigations beyond the city's standard exactions. If Barbara Plaza were proposed in Santa Monica it would have been approved "over the counter" because it was beneath the size threshold for CEQA review. Riverside Arms would have been reviewed (because of the zone change, not the project size) but then probably would have received a negative declaration in Santa Monica.

Santa Barbara required CEQA review of its own routine project, just as it would have of the other two routine developments.[17] After review, little Barbara Plaza was given a negative declaration but with conditions pertaining to traffic mitigations. Riverside Arms would have been reviewed because of its size and would have been subject to tough traffic mitigation measures, depending on its exact location. Monica Tower, on the other hand, if it had been approved in Santa Barbara (which it would not have been), would have required a full EIR and extensive mitigations.

Even when CEQA was invoked, the local implementation differed, as did all the other regulatory mechanisms we have discussed. For example, when Riverside officials rezoned 150 acres to accommodate the elusive auto mall described above, the city, though requiring certain traffic improvements, issued a negative declaration. They reasoned that an auto mall would not be any worse environmentally than the industrial uses already allowed on that site. Even though the environmental impact of existing industrial zoning was never evaluated, it was taken as the ground floor for loosely judging the impact of alternate uses. In contrast, Santa Barbara required significant mitigations for its Rancho***Barbara auto center. For such a large project in Santa Monica, mitigations would likely be secured through a development agreement.

The reliability and usefulness of EIRs also varied. Until 1990, Riverside developers were allowed to hire their own consultants to do the reports, with the city providing oversight. Santa Barbara and Santa Monica had long severed this patron/client tie between developers and experts by directly contracting consultants to prepare EIRs (at the developers' expense). Riverside changed its approach only under pressure from citizens who opposed Riverside***Place and threatened to sue the city, citing a Los Angeles County court case where environmental documents prepared under contract with the developer were nullified.[18]

Another variation in environmental review was the way government notified citizens of pending development decisions. CEQA review requires that neighbors affected directly by a project be notified of all public hearings, either by posting signs at the building site, by giving them written notice (by flier or mail), or by publishing newspaper announcements. Since routine projects in Riverside received no public review, there was virtually no occasion for notification; the first anyone would know of a project would be when construction began. Santa Monica applied EIR processes to a larger proportion of projects than Riverside, using both mail and newspaper publication, with selective posting. Santa Barbara exceeded the basic CEQA requirements and the other locales, using all three methods of public notice as a matter of course.

Probably more important than methods of notifying neighbors was the kind of treatment media gave to the substance of EIRs. Our cities differed, even though all their major papers were pro-development. The Santa Barbara newspaper gave consistent and detailed coverage to the content of environmental studies, specifying such matters as the number of car trips a project would generate, effects on air quality, and expected congestion at particular intersections. The paper advised its readers to inspect the EIRs, in one case editorializing that an EIR for a controversial downtown department store project (which it strongly supported) contained "information vital to a decision on the ... issue" (*Santa Barbara News Press*, October 9, 1983: D1). On this project alone, the paper provided information on the EIR content (much of it damaging to the project) in fourteen separate stories and also indicated time, date, and place of upcoming hearings and meetings related to the development. These stories were all played prominently; three of them (two of which focused solely on the EIR) were front page, in one case just below the day's lead story: "U.S. Troops Preparing for Grenada Withdrawal" (*Santa Barbara News Press*, November 3, 1983: A1). Five other articles appeared on the front page of other sections.

In regard to an equally contentious and large-scale Santa Monica project (a proposed beach hotel that was eventually defeated at the polls), only two articles in the *Outlook* made direct mention of the project EIR's content. In one of these instances the paper merely paraphrased the project opponents' use of the EIR's findings on air quality impacts.

The *Riverside Press Enterprise*'s coverage of the city's best-case project barely mentioned the project's EIR (one of the relatively rare instances in which there was one in Riverside). Of twenty-two news articles on the project, only one even mentioned the existence of an EIR (*Riverside Press Enterprise*, April 3, 1991: B3), saying that the city council would consider the final compromise on the project at a June 4, 1991, meeting, "after development plans and an environmental impact report are revised." An earlier article mentioned "a consultant's study" on traffic impacts, but this appeared in connection with one of the developer's justifications for higher densities, rather than as an element of environmental review. This coverage by the Riverside paper was not due to some thoroughgoing lack of fairness toward environmental review; indeed we have documented elsewhere that when it came to election contests the Riverside paper showed less evidence of systematic pro-growth news bias than the other two papers (Warner and Molotch, 1992: 101, 102). Rather, the coverage suggests a difference in local planning cultures and practices.

Our comparison tells us that Santa Barbara's coverage alone was extensive and informative, albeit sometimes critical of what the paper regarded as overly strict methodologies for assessing impacts. Nevertheless, its extensive news treatment surely extended consideration of substantive environmental information to a broader public. Santa Barbara news coverage likely helped make the substance of EIRs an intrinsic part of public perception and action.

Local Knowledge Base for Planning Decisions

Effective growth control and meaningful environmental review depend in part on the quality of local planning knowledge. Otherwise, there are few benchmarks for judging particular project proposals and the associated environmental review documents. To know there will be a given impact, people in a locality must first understand such things as the infrastructure holding capacities, the nature of cumulative impacts, and the fiscal and environmental costs and benefits of adding to the built environment. No single study can do this; rather, a base of knowledge must already be in place. Our cities were different in the degree to which they had this information at hand.

Santa Monica had made serious efforts to determine the fiscal impacts of growth by land-use type. In 1982, the city assembled information on the fiscal costs and benefits of office uses in the city. Consultants calculated the revenues generated by offices and compared this to the public costs (including general fund expenditures) attributable to these land-uses (Hamilton, Rabinovitz, and Szanton, Inc., 1982). They figured office workers (and clientele) as one portion of the total number of people using everyday city services and facilities (e.g., streets, parks, police protection). This fraction of the total public costs was assigned to office uses and compared to revenue generated by office buildings (using a random sample of buildings stratified by building size). At the same time, surveys of office employees were used to estimate the affordable housing demand generated by office uses. Based on the assembled information, Santa Monica's parks/housing fee for office development was assessed to offset the proportionate costs to the city of this form of growth. In 1990 the city calculated the net public costs of all non-residential development, using the 1982 study as base information, and found that, with the exception of hotels, commercial developments did not generate revenues beyond their costs to the city (Economic Research Associates, 1990). Development opponents frequently cited chapter and verse from the city's own studies and database to argue against further commercial development and public/private ventures.

Santa Barbara seemed to have even more complete self-knowledge. Using the information assembled through past planning efforts (including a study by the consultants who assessed fiscal impacts for Santa Monica—see Economic Research Associates, 1986), Santa Barbara planners were able to judge projects in terms of their impacts on carrying capacities and the types of mitigations that might offset negative effects. Because they knew a project's prima facie impacts, city officials were able to know in advance what kinds of impact reports (containing what sorts of specific information) to require. Political leaders might not always make use of the best information available, but planning information was a resource that could be mobilized by growth opponents, just as it was a constant source of information for project sponsors. Santa Barbara's effort to link planning analysis, information bases, and long-term policy to population goals rendered planning information a more potent political resource. When Santa Barbara planned for the future, it planned with an end state in mind (a given level of population, types of physical configurations, and level of infrastructure quality) and could evaluate any project in terms of its effects on that end state.

Riverside had a weaker base of planning knowledge. Information on growth impacts was focused narrowly and prepared mostly under the patronage of project sponsors. Public leaders still assumed, without evidence, that increased commercial growth would solve their "growth problem" of too much housing and too few jobs, rather than aggravate it by generating still more of both. In the words of the Riverside planner, "There is just a general assumption that office, commercial, and industrial are good."[19] Even the slow-growth officials supported new commercial/industrial development without knowing whether it solved or created fiscal (or other) difficulties.

Resisting Growth: A Matter of Degree

Projects that were approved and built in the three study areas reflected, in physical form, the degree of growth coalition influence. Riverside planners might have liked to do more but ended up accepting most of what developers proposed—including the size, density, and design—on all but the exceptional project. Little was done to extend environmental review beyond state mandated requirements or to enhance citizen involvement.

Santa Barbara, on the other end of the scale, set the tightest physical limits on developments and extended environmental review to even the smallest commercial projects. Santa Monica permitted much more physical development but steered projects strongly toward community planning and "revitalization" goals. Within this populist political context, citizen involvement was promoted heavily (including city funding of land-use watchdog organizations), but environmental review was less universally applied than in Santa Barbara.

The different local planning contexts shaped the negotiating stance of city planners vis-à-vis developers, as we see in the Santa Barbara and Riverside responses to proposed auto malls. Santa Barbara officials saw good reasons to approve (and even promote) an auto mall at that site but were unwilling to concede planning standards or other community goals (e.g., affordable housing) in order to get it. Riverside officials, on the other hand, were so anxious to land another auto mall that they rushed through a full package of concessions for the developer. When planners had the political and administrative support to say "no," as in Santa Barbara, they were in a position to force negotiation. Otherwise, as in the Riverside example, public concessions were made up-front, leaving developers maximum flexibility over the content of their projects, with little or no accountability for negative impacts and public costs. Ironically, this situation seemed to provide a context in which projects were more likely to be approved but were then left unbuilt—financial gains may have been made by speculating on property values, but profits were not produced by physically developing the land.

On the whole, however, and compared to past eras, growth interests did not go unchallenged in any of our study sites. Property rights were still accorded deference in all places, but officials included a wider range of constituencies and issues in the process than they would have in any of these cities in the pre-environmental era. Still, the level of challenge was, as we now have documented, very different across our three "growth-control" cities. In the next chapter we will lay out the results in terms of specific sorts of public benefits gained by each approach to regulating development.

Notes

1. For examples of past surveys, see Association of Bay Area Governments, 1980, 1984; Construction Industry Federation, 1988; Cervero, 1988; Frank and Downing 1988; Kaiser, Burby, and Moreau 1988.

2. Interview 015. For explanation of interviews, see Chapter 2.

3. The actual projects are: Santa Barbara, 801 Chapala Street, mixed-use office/retail; Santa Monica, 1250 Fourth Street, mixed-use office/retail; Riverside, Spruce Apartments, 1000 Spruce Street.

4. We excluded redevelopment or revitalization projects because the proactive government involvement muddles the distinction between developer and regulator.

5. The actual best case projects are: Santa Barbara's "Rancho Arroyo" mixed use residential/auto mall, Hope Avenue and Hitchcock Way; Santa Monica's "Water Garden" mixed use office-retail, bordered by 26th Street, Colorado, Cloverfield, and Olympic; Riverside's "La Sierra" Mixed Use Development, Indiana & La Sierra Streets. The "La Sierra" project ended up not being built under the conditions specified in the initial planning documents because of unanticipated neighborhood protest, circumstances we describe in the text below.

6. Interview 034.

7. Interview 015.

8. Santa Barbara also allows even higher densities (up to 27 units/acre) for studio apartments in the "transitional" zones around downtown to encourage housing there, another planning goal.

9. Interview 015.

10. In Riverside, rezoning to allow apartments is what generates most opposition (see below).

11. The Riverside planner indicated that Riverside would also not rule out higher housing densities, if proposed by the developer.

12. Interview 036.

13. Interview 040.

14. Interview 040.

15. For description and analysis, see Office of Planning and Research, 1986; Duggan, Moose, and Thomas, 1988.

16. Interview 082.

17. The rules in surrounding Santa Barbara were also stringent. In a comparison of fourteen cases throughout California, researchers found that Santa Barbara County required EIRs on the highest proportion of projects (Landis, Pendall, Olshansky, and Huang, 1995).

18. See Torres, 1990, for coverage of the Los Angeles case.

19. Interview 040.

6

Indirect Effects:
How Building Rules
Make Growth Different

The rules developers must navigate to get their projects approved clearly vary across our cases. This variation affects the type and size of developments that authorities allow. Different approaches to growth control guide developers toward satisfying particular local planning goals; we realized this as planners gave examples of the kinds of projects they would approve (and had already approved). Here we systematically compare the relationships between building rules and the achievement of larger policy goals.

Growth regulations provide local regulators with opportunities to control the kinds of spillovers that most concern them—or, to take the mirror image, to engender the kinds of communities they most wish to sustain. Recall the local planning "agendas" we described in our presentation of the study sites (Chapter 2). Santa Monica placed high priority on social equity but also on the scale of development and on environmental impacts. Santa Barbara's central concern was the texture of the physical city and its natural environment, with secondary (but still substantial) attention to social equity. Riverside aimed to build the local economy and job base, with a lesser, but still present, interest in environmental preservation.

Although we rejected the idea of using a single measure for the degree of growth control (as explained in Chapter 3), we found it useful to compare the implementation of particular types of regulation—like the fees charged for traffic mitigation or affordable housing contributions required of commercial development projects. We complement our qualitative comparisons by contrasting the fees that were collected for particular purposes: mitigating traffic, preserving natural environments, maintaining affordable housing, providing social equity, improving urban spatial form, and enhancing project aesthetics. Again, we use the six project proposals as concrete reference points to compare the projects' public benefits.

Traffic

Traffic shapes the everyday experience of getting around and being in a place, affecting safety and giving localities their many variations of small-town, big-city, or suburban feel. Managing these circulatory systems also affects land-use patterns, including the orientation and design of all other land uses. Road access distributes rents and nuisances in modern cities. At an ecological level, the astonishing amount of space— often close to one-third—devoted to paved streets, sidewalks, driveways, garages, and interchanges impacts ground water quality, the recharging of aquifers, erosion, and coastal outfalls, as well as wildlife habitats.

In fiscal terms, solving traffic overloads can be extremely expensive, since solutions are usually implemented after the fact. Cities may be faced with buying developed properties at prohibitive costs to expand old roadways or build new ones. The alternative is to limit traffic demand (a common and powerful impetus for growth controls) or arrange land uses to reduce transportation needs or at least minimize chaotic outcomes. A simple requirement that a road entry be moved from one side of a project to another can have significant preventive impacts. Gaining initial control over how much traffic a project will generate can minimize a wide variety of impacts.

At the broadest level, each city had some form of policy aimed at controlling traffic effects. Santa Barbara's transportation requirements were the tightest and were applied most consistently. The city charged developers $2,000 for each peak-hour trip generated by even routine projects, at least in the downtown area. Riverside never charged this type of fee under any circumstance. Santa Monica had no standard traffic fee and often levied none, even on larger projects. For example, neither the routine Santa Monica project (which included 321 parking spaces) nor a hotel with 221 parking spaces that was approved the same year required any traffic fees. But the city did levy a hefty $3,000 per generated trip on Monica***Centre, its best-case project. The amount of parking provided in this project (see Photo on page 106) gives one a feel for the traffic it generated.

Planners in virtually every jurisdiction we studied regarded traffic as a major problem affecting local residents. In reviewing each of their best-case projects, planners in our three principal cities weighed traffic impacts heavily. Santa Barbara claimed to have improved traffic conditions by allowing the Rancho***Barbara project because the required road improvements enhanced the "level of service" at surrounding intersections;

Underground parking at Monica***Centre

in their best-case projects, Santa Monica and Riverside both secured more traffic improvements or fees than ever before. The Santa Barbara and Riverside project approvals called for the developer to make about $3 million in direct traffic improvements. In Riverside this amounted to one-third of the total costs for needed improvements, whereas the Santa Barbara developer shouldered all costs. In other words, Riverside developers paid $4.00 per square foot of commercial/retail floor space, compared to the whopping $92.00 per square foot paid by developers in Santa Barbara.[1] The developer of the Santa Monica best-case project paid $6.4 million up-front for traffic improvements—$5.00 per square foot of floor area, an amount augmented, on this particular project, by a $200,000 annual transportation-demand management fee. Still, even after thirty years the total traffic mitigations per square foot of floor area would be lower in Santa Monica than in Santa Barbara.

On the bigger projects, Santa Barbara officials collected strikingly more at the outset for traffic improvements than either of the other study sites. But they also required traffic mitigations on small projects like Barbara Place (including free employee bus passes and showers for those commuting by foot, bike, roller-blade, or other form of self-propulsion).

The three cities also differed in their follow-through on traffic reductions. On its best-case project, Riverside imposed a laundry list of recommended mitigations (e.g., van pool incentives, preferential parking for ridesharers, bikeways, and bike parking) but established no mechanism to assure that any of them would be implemented. Monica***Centre, on the other hand, offered an incentive for follow-through: Any traffic reductions accomplished after construction would be counted directly as a credit against the annual traffic fee. This was an exceptional case for Santa Monica, however. Up to 1990, the city had done less than Santa Barbara to evaluate traffic impacts comprehensively and to mitigate new development. Follow-up in Santa Monica and in Santa Barbara, although certainly more consistent than in Riverside, was problematic, in that even a substantial city staff and watch-dog citizenry could not make sure that every stipulation was obeyed. The Santa Barbara County auto dealership fiasco (across the city line in nearby Goleta) illustrates a larger class of follow-up difficulties (see Chapter 4).

In Santa Barbara, any project that sent as much as one new peak-hour auto trip through an "impacted intersection" (one already operating close to capacity) had to provide mitigation. Public officials would not allow much widening of roadways (because of issues of community character)—even if developers were willing to pay for them. Limits on road expansion, thus, became a potentially effective growth limit, independent of the land-use restrictions attached to any given parcel. Although we found no evidence that traffic design had reduced aggregate levels of development, we observed instances where traffic standards guided projects toward one site over another, thus altering the geography of development. This was clearly a stronger possibility in Santa Barbara than in the other two jurisdictions; and when it occurred, it was regarded by planners as a superior method compared to requiring complex mitigations at a bad site.

Natural Limits

A lack of water is the chronic problem for Southern California development, solved over the years only through extraordinary feats of engineering and politics, and by grossly disturbing natural landscapes, near and far. Santa Barbara and Santa Monica commercial developers had to offset new demands for water by achieving conservation at the site or elsewhere in the city. Santa Monica introduced this requirement because of limited wastewater capacity, whereas Santa Barbara was conserving local water supplies. The five-year California drought of the late 1980s brought more stringent standards; both cities required of new development that it produce water savings equal to half the previous on-site water use. To appreciate this in real terms, developers of a 100,000-square-foot office building in Santa Monica had to retrofit an elementary school with more water-efficient toilets and plumbing at a cost of about $35,000.

Our "project-peddling" with planners did not include a detailed analysis of the natural characteristics of each site or of how approvals dealt with environmental spillovers, a fact that limits our comparisons. Some of the environmental requirements made in Santa Barbara—for example, the protection of creek drainage at Rancho***Barbara—may have been due to unique characteristics of a particular site more than to a difference between localities.

Still, we can draw some conclusions about how our study sites managed natural impacts. Allowing only smaller scale development, as in Santa Barbara, creates less traffic congestion, which has positive air quality outcomes. Santa Monica's restrictions on wastewater production helped stave off expenditures for new groundwater control systems, while also limiting, to some degree, pollution in Santa Monica Bay. Santa Barbara found some ways to reduce solid waste—for example, by requiring that the local newspaper facility use recycled pulp in its daily editions as a condition for expansion of its production. Finally, policies governing landscaping, revegetation, and grading had direct consequences for plants and wildlife. Santa Barbara County required developers not just to plant something but to preserve the mix of local vegetation. This requirement helped sustain existing ecosystems by blocking exotic species and fostering critical masses of native habitats. Such requirements can also give wildlife corridors some chance of remaining intact—all of these considerations were often on the table in Santa Barbara (both city and county) negotiations with developers.

Affordable Housing

In terms of housing aid, no exactions from new development in Santa Monica could rival the massive effects of rent control. Nevertheless, Santa Monica used the development process to secure still more affordable housing. Characteristic of Santa Monica's political style, a planner put it bluntly: "Providing more housing is much more important than providing new office space for accountants or somebody."[2]

Santa Barbara's housing exactions rivaled Santa Monica's in ways we did not expect. Indeed, it is something of a challenge to determine which city had the toughest regulations (putting aside Santa Monica's rent control law). Both cities had established methods to determine how much developers should contribute toward affordable housing. Santa Monica used a fee per square foot on office projects or as a set portion of new residential units. In Santa Barbara, planners calculated the increased demand for affordable housing caused by each commercial or industrial project (residential projects had no affordable housing requirements unless the developer wanted bonus densities).

Using the Monica Tower project as a basis for comparison, Santa Monica collected $1.63 per square foot for housing, a total of $152,325.[3] Santa Barbara planners estimated such a project would generate demand for thirty-six affordable housing units, requiring a housing payment of $1.4 million, or $15 per square foot (amounting to a $39,000 subsidy per housing unit), making the Santa Monica fee appear very low. Rather than paying fees of this kind, Santa Barbara developers had elected the other options of building affordable housing on-site or off-site, or of subsidizing existing housing to an affordable level. In the two years the housing fee had been in effect (since 1988) no builders had actually paid the in-lieu fee for affordable housing—choosing to provide comparable housing in some other way.

Another way to compare the gains in lower cost housing is to figure the rent or sales revenue that developers lose over the long term in meeting housing requirements. Santa Barbara developers had to guarantee that rents would be affordable to moderate income families for seven years—for example, a family earning 80 percent of median income would spend a maximum of 30 percent of their earnings on rent.[4] Santa Monica required that housing remain affordable to moderate income families for a period of fifty-five years. Because of the longer term, Santa Monica's requirements seem more stringent than Santa Barbara's, but Santa Monica required a much lower number of on-site affordable units.[5]

The number of affordable units and the term of affordability both figured into the cost of housing mitigations in "lost rent."

For rents to meet affordability standards in Santa Barbara in 1990 they could not exceed $862 per month. The "lost rent" would then be the difference between this figure and market rent. Since data on market rent for these units were not available, we estimated conservatively the total subsidy that developers put into affordable housing. For this exercise, we assumed that market rents equaled 30 percent of the earnings of a family at median income (probably an underestimate given the amount families spent on rents in the late 1980s). The "lost rent" was then the difference between affordable rent for a median income household (30 percent of median income) and affordable rent for a moderate income household (30 percent of moderate income, where moderate income is 80 percent of median).[6]

Using the "lost rent" formula, we compared the housing subsidies in Santa Barbara and Santa Monica based on the case of Monica Tower. Under Santa Barbara review, the project developers would have had to provide thirty-six affordable housing units for seven years, entailing a subsidy of $651,672. Had Santa Monica required developers to build affordable housing rather than pay the housing fee, they would have asked for five affordable units for fifty-five years. The subsidy would have equaled $513,480. This amount far exceeds the $152,325 they did collect but still falls short of the Santa Barbara subsidy. Because developers in Santa Monica had a clear financial incentive to pay the housing fee rather than build affordable units, the Santa Barbara housing exactions turn out to be far more stringent.[7]

For every square foot of commercial development allowed, Santa Barbara demanded more investment in affordable housing. But in absolute terms, one might expect Santa Monica to lead in affordable housing gains since it had more techniques for generating affordable housing through development approvals, including inclusionary zoning (stipulating that all residential projects must include affordable units), bonus densities, office mitigation requirements, and development agreements.[8] Santa Barbara did not have inclusionary zoning, and housing mitigation for commercial development was not required until 1988. But it turned out that the variety of exactions did not necessarily predict the amount of affordable housing produced. Through 1990, Santa Monica had gained seventy-six affordable housing units through inclusionary zoning requirements. The city did not track the amount of affordable housing gained through bonus density allowances, but housing staff estimated a total of no more than fifty units through 1990. During the same period, Santa Monica collected more than $11.26 million from a combination of office mitigation fees, in-lieu fees for inclusionary zoning, and housing exactions included

Housing construction in Santa Barbara

in large-scale development agreements. By 1990, 162 units of affordable housing had been produced with the help of these funds (leaving a balance of approximately $5.2 million). At that rate of subsidy, the city could potentially assist another 140 units.

In Santa Barbara, the housing mitigation program had produced meager results, only seven units by 1990. But 354 bonus density units had been built. The total of all the exaction programs suggested that more housing was on-line in Santa Barbara than in Santa Monica (361 versus 238 housing units). If we included Santa Monica's surplus housing funds, Santa Barbara fell somewhat behind, but this comparison did not account for the subsidized units in the pipeline there, which would again place it in the lead. If we were to add further refinements to reflect Santa Barbara's smaller population size, lower levels of poverty, and more modest commercial expansion, its relative performance would appear still stronger (again, however, omitting consideration of the crucially important Santa Monica rent control program). Although Santa Barbara may have constrained new housing supplies through its zoning and other control regulations, it set the pace in producing affordable housing.[9]

In addition to adding affordable housing stock, Santa Barbara and Santa Monica have guarded existing supplies. In both cities, developers were required to replace any affordable housing units that they might destroy. Santa Monica extended the same logic to low-cost hotel lodgings as a way to protect single-occupancy dwellings that sheltered people who otherwise might be homeless.[10]

In Riverside, neither commercial nor residential developers were required to provide housing mitigation. One might argue that this was justified because the private market provided lower housing costs relative to the Los Angeles region. But rents that seemed low from a regional perspective were beyond the reach of lower income Riversiders because their incomes were also lower. In 1990, a moderate-income family of two could afford only $433 per month, compared to the $545 per-month market rent asked at the routine project, Riverside Arms. For the low-income segment of the community, earning less than 50 percent of median income (the population targeted for affordable housing in Santa Monica beginning in 1990), the rents at Riverside Arms would have to have been slashed to $271 per month. In relative terms, growth in Riverside was not addressing unmet local housing needs, though it may have provided some needed supplies for the broader region.

Social Equity

Santa Monica's emphasis on social equity took the form of a wide array of "linkages" in which developers were asked to deal with social problems

*On-site child care at Monica***Centre*

that only indirectly, at best, were created by the intrusion of their projects. This required not only a particular brand of local political will but also a good amount of legal finesse that might not have occurred in places with less ingenious and tenacious leadership.

The city could not "take" anything from developers without facing litigation. The key test of exactions under the taking doctrine was whether or not a city's demands were linked directly to problems caused by the project itself. There had to be a "nexus," in the legal language, between a problem created by the project and the mitigation that was required. The nature of local social problems, as opposed to issues like traffic creation, made them hard to pin to any specific project, or indeed to any projects at all. Sometimes, as when a given development eliminated poor people's housing, a case could be made that aid to the homeless should be provided to deal with the potential effects of the development. But to provide still additional aid to the homeless when the project did not destroy housing was quite another matter.

Santa Monica bypassed the nexus test by using developer agreements, transforming an instrument commonly used to buffer developers from future environmental regulation or political interference into a tool for planning innovation. Under such agreements, specific conditions (which can be more or less stringent than prevailing policy) are worked out with the developer on a case-by-case basis. For Monica***Centre, the developer agreed to put up $300,000 for homeless services and $150,000 for public art and on-site child care (see Photo on page 113). In other projects, Santa Monica required job training and affirmative action programs. One prospective developer responded to neighborhood demands by offering a straight 6 percent of the $43 million dollar project costs to the local neighborhood group to use as it saw fit (Fulton, 1985: 88).

Developers went along with such exactions, quite aware of the detoured nexus test, when they believed that their projects were still profitable. Some resented the method, but strident opposition and the legal hassles were not worth a fight—so long as aggregate costs were reasonable, given potential returns.

Such social welfare gains were not a standard feature of building rules in other localities, not just because of the legal issues but also because the necessary underlying political pressures were just not strong enough in any place besides Santa Monica. The larger ideological structures in play in this city during our study period allowed a new arena—the development process—to broaden toward social welfare issues. Development became another site of the city's distinctive social welfare stance, evidenced as well by its program of free daily meals on the front lawn of city hall and by an unwillingness to prosecute people for sleeping outside. Although the city's policies were not accepted by all segments of the community, and although

Homeless man in front of Monica Tower

they were altered substantially in later years, Santa Monica had come to an uneasy peace with its visible homeless population (see Photo on page 115).

In Santa Barbara, social missions were never as important a part of city programs as they became in Santa Monica. The community was divided on aid to the homeless. Some perceived the highly visible homeless population as a visual blight on the handsomely tended environment and as a rude detraction from the tourist economy. The city approved oceanfront developments that wiped out the town's established "hobo jungle" and in one instance wantonly cleared vegetation that otherwise shielded homeless encampments from public view and official surveillance. A 1990 city ordinance prohibited sleeping in public places. The ordinance was aimed at the homeless and drew nationwide publicity, including a running commentary by Gary Trudeau in his syndicated *Doonesbury* comic strip. It also generated local demonstrations led by national advocates of the homeless, further complicating local political dynamics on the issue. Even so, advocates for the city's affordable housing programs used the prospect of still more homeless as a talking point to safeguard and expand mitigations for any redevelopment effort that might destroy inexpensive units. Requiring direct aid to homeless people from developers, however, was not in the realm of political possibility.

Even on less controversial social issues like child care, Santa Barbara was again less motivated than Santa Monica, partly because its leaders did not want to be seen as sponsoring development specifically to gain the social payoffs. Commenting on Santa Barbara's only development agreement, our planner informant said: "Planning commissioners weren't interested in pursuing this approach on other projects; it was a little too much like being in bed with the developer, approving something because of the money the city would make. The city is pretty comfortable working with either the nexus they can establish on project impacts or the regular budget process for improvements."[11] From an environmentalist perspective, taking cash to allow development was sometimes considered a "sellout"—whether to gain money for social causes or any other sort of community goal (e.g., aid to the art museum or symphony).

Riverside, although arguably in a region with the most severe environmental difficulties and with social problems of poverty and high welfare needs, had an even narrower range of exactions than Santa Barbara. At least in part this lack of exactions was due to the different political priorities of its leaders. But it was also due to public officials' sense that they could never get away with asking for much, lest developers simply abandon their proposals. As the Riverside planner commented when looking over the case studies: "It looks like people would give their grandmother away to get something approved in Santa Monica.... The traffic-demand management fee, art fee, homeless fee, these would all be wonderful to try to get, but we would never get them out here. The developers would probably turn away and say, 'You are kidding.'"

Urban Spatial Form

Growth controls can shape the particular configuration of the city—for example, by centralizing certain functions or dispersing others, by building a critical mass of a particular type of urban agglomeration or serving to discourage it. In Santa Monica, regulations that discouraged office development at the expense of housing, particularly along the oceanfront, stimulated office construction in the areas planners favored for the purpose (the eastern zone, made up predominately of warehouses and otherwise underutilized acreage). Developers built not only offices in this zone (see Photo on page 118) but also an entirely new kind of city district with related support businesses (office equipment, restaurants, etc.).

In Santa Barbara, the only way to make it through the increasingly tight commercial development rules was to construct low-profile structures, as exemplified by Barbara Plaza. The effect was to fill in the urban fabric rather than to spread it outward or to cause high-rise "bumps" on the horizon or traffic hot spots in the road system. Barbara Plaza made partial re-use of an existing but underutilized structure, as per city preferences for rehabilitation of such buildings. The photo on page 119 shows another example of this type of infill on Santa Barbara's main downtown street.

The process for gaining exemptions under Santa Barbara's 1989 commercial cap also had the potential to further city goals. Under the rules, developers could replace any size structure with one of the same size (plus 3,000 square feet allowed as an addition to any building). This rule increased the attractiveness of replacing marginal structures, especially in the valuable downtown core (though the tough landmarks protection system preserves old buildings of special merit). The commercial redeveloper could also tap the water that had been used historically on that site, thus getting around the city's ban on new water hookups. A Santa Barbara planner gave an example of the good fortune of one property owner who owned a large, industrial, airport-hangar-type building: "They have a significant amount of square footage, which was a detriment to that property two years ago. I talked to the guy today, and he has had people coming out of the woodwork asking how much he wants for the property, and everybody he talks to, it just keeps going up."[12] This combination of incentives reordered land markets in Santa Barbara by favoring existing density patterns across the city and, especially, downtown redevelopment, a long-term planning aim. Of course, this development conflicted somewhat with the overall goal of limiting

Santa Monica office construction

Santa Barbara infill development

growth and further helps explain the otherwise curious finding of so much growth under growth control (see Chapter 4).

Project Aesthetics

Architecture, landscape, and other aspects of the built environment help define the character of a community. Every building, every tree, every sign has meaningful implications for character of place. Improving the aesthetics of a place may be a goal in itself, but such improvements have much broader local impacts, including the kind of economy a place can sustain (Lash and Urry, 1994; Zukin, 1996; Molotch, 1995). Localities with high aesthetic resources have better opportunities to create tourist and retirement economies, particularly affluent ones. High aesthetic accomplishments in the public environment attract cutting-edge, high value-added industries from, for example, the development, design, and finance sectors. Good-looking public environments, as defined by those with the resources to take themselves and their industries to places that please them, attract individuals with advanced tastes who then construct and demand other facilities (e.g., private homes, restaurants, boutiques) that further reinforce the "virtuous circle" of high amenities, high private expenditures, and benign fiscal effect. Not only individuals have cultural capital; so do places, and with substantial consequence.

Of all our study sites, Santa Barbara had the kind of economy most driven by aesthetics—tourism and high-end retirement, as well as research and development. It also had the longest history of aesthetic activism and design regulation. This is not to say that economic interests singularly determined local culture. Indeed, early efforts to protect the beauty of Santa Barbara by citizens who had made vast fortunes elsewhere created the opportunity for an aesthetics-based economy for later generations of fortune builders.

The city required architectural review for every structure (it had, recall, the country's first such architectural review board), down to the color schemes for exterior paint; paint chips had to be submitted with architectural plans. Many projects were deemed to have historical or architectural impacts, especially those built in the downtown core, and therefore had to go before the city's official Landmarks Commission, which provided a second level of design review, where many cities have none. All downtown projects must adhere strictly to the Mission Revival architectural theme that has been etched into Santa Barbara's urban form for generations (see Photo on page 121); this theme includes red tile roofs (an expensive material) and, at times, extensive exterior use of wrought iron trellis and railing material (another somewhat costly element). Outside the central core, the city followed different, but still relatively tight, aesthetic guidelines.

Downtown Santa Barbara architectural style

Design review in Riverside and Santa Monica was generally more concerned with how a project "fit" within the immediate neighborhood; planners left such things as architectural style to the developers' discretion. In Santa Monica, scale and compatibility with immediate surroundings loom large in the approval process, often because of neighbor's fears that a project might block views or sunlight, or clash with prevailing neighborhood design. Standards for Santa Monica's Ocean Park neighborhood encouraged new development to respect the "underlying patterns and construction techniques of diverse existing architectural styles: craftsman bungalow, Spanish colonial revival/Mediterranean, and international style/modern (Sedway Cook Associates, 1989: IV–5). But even here, and certainly in Riverside, a far smaller proportion of projects go through architectural or landmark review than in Santa Barbara.[13]

Exactions on the Ground: A Tally Between Places

We have emphasized that the simple presence of growth control measures on the books is at best a crude indicator of what goes on at ground level. Similarly, counting numbers of controls in place, including the numbers or types of fees collected, leaves out a lot of important information, such as the actual content and application of such fees and how they are triggered and calculated. High fees may indeed stymie growth, or they may instead deflect development to other parts of the city with lower fees, or they may even accelerate growth by providing critical public infrastructure. High fees may create standardized costs in ways that spare developers more costly and contentious ad hoc assessments of specific project impacts.

With all our caveats in mind, our project peddling gives us a way to compare places systematically in terms of the fee structures that count most—the ones actually exercised in particular situations. We now offer a summary, based on our project case studies, of how exactions are used in practice. Table 6.1 summarizes the number and the type—traffic, housing, parks, and social services—of exactions that would be applied for the case studies in each locale. We do not include improvements at the site or in the immediate vicinity, such as upgrading adjacent intersections or viaducts, or dedicating land for parks, fire stations, or schools. We also exclude school fees from this comparison because they were mandated in specific ways by state laws.

A place that imposed every type of fee exaction on every project would have a total of twenty-four exactions (four types of exactions on each of the six projects). We see that Santa Monica would have collected the greatest number of fees: sixteen out of a possible twenty-four (67 percent), versus eleven that would have been collected in Santa Barbara (46

TABLE 6.1 Case-by-Case Use of Exactions

	Traffic	Housing	Parks	Social Services
Santa Barbara				
Barbara Plaza	yes	no	no	no
Monica Tower	yes[a]	yes	no	no
Riverside Arms	yes[a]	yes	no	no
Rancho***Barbara	yes	yes	no	no
Monica***Centre	yes	yes	no	no
Riverside***Place	yes	yes	no	no
Santa Monica				
Barbara Plaza	no	no	no	no
Monica Tower	no	yes	yes	no
Riverside Arms	no	yes	yes	no
Rancho***Barbara	yes[b]	yes	yes	yes
Monica***Centre	yes	yes	yes	yes
Riverside***Place	yes[b]	yes	yes	yes
Riverside				
Barbara Plaza	no	no	yes	no
Monica Tower	no	no	yes	no
Riverside Arms	no	no	yes	no
Rancho***Barbara	no	no	yes	no
Monica***Centre	no	no	yes	no
Riverside***Place	no	no	yes	no

[a] in designated traffic impact area

[b] if approved by developer agreement

TABLE 6.2 Use of Exactions by Project and by Type

	Santa Barbara	Santa Monica	Riverside
Land Use			
Residential	50%	50%	25%
Riverside Arms			
Office/Retail	33%	40%	20%
Barbara Plaza			
Monica Tower			
Monica***Centre			
Office/Retail/Residential	50%	100%	25%
Rancho***Barbara			
Riverside***Place			
Routine versus Best Case			
Routine Projects	33%	27%	20%
Best Case Projects	40%	100%	20%
Exactions by Type			
Traffic	100%	50%	0%
Housing	83%	83%	0%
Parks	0%	83%	100%
Social Welfare	0%	50%	0%

percent) and six in Riverside (25 percent). We can analyze these exactions further, in terms of project type, routine versus best cases, and exaction type, as in Table 6.2.

Park fees were the only type of consistent exaction that applied in Riverside but not in either of the other cities; Riverside collected park fees on all new construction projects. In this regard Riverside seemed to surpass Santa Barbara, which does not collect the standard park fee authorized by state law. However, as a Santa Barbara planner explained, it is unlikely that Riverside had a higher priority for parks than Santa Barbara. Rather, the state limits the use of these funds to land acquisition (versus park development or maintenance). Since the city of Santa Barbara already owned thousands of acres of undeveloped park land (acquired in part through land gifts to the city), the city did not collect the state authorized fee, which only would have added to the task of developing and maintaining the lands they owned.

Compared to Riverside, the other two cities imposed a far greater variety of exactions. Santa Barbara collected the most exactions on routine projects, always assessing developers for off-site traffic impacts. But Santa

Monica was the only city that collected every kind of exaction on at least some of the cases, and it was also the only city to fund specific social services from development fees. Santa Monica collected especially on best-case projects.

Although fee differences conform somewhat to expectations, some things are "off," such as Riverside's park fees. Nor do fee schedules provide a clear understanding of differences between Santa Monica (which collected the largest number of fees) and Santa Barbara (which collected them on a wider range of projects). Most important, and returning to our larger theme, any formalized method of fee counting does not capture the far-reaching changes in development review practices that can incorporate regulation into the very design, content, and scale of development.

In Santa Barbara and Santa Monica, land policy and rule implementation gained a certain autonomy from market impulse and became an arena for more conscious and participatory decisionmaking. A general policy, such as Santa Barbara's commitment to neighborhood compatibility, was invoked rather consistently to produce conforming building usages, height limitations, offsets, and aesthetics. A Santa Barbara planner compared the city with other places he had worked: "I have seen a greater tendency here to want to live within the regulations. Even down to signs.... This community takes every square inch of sign surface seriously." Through the overall politics in place and the "wall of no" that developers might face, Santa Barbara and, to some extent, Santa Monica were able to advance their own planning goals through the administration of project approvals. Building rules helped Santa Barbara push environmental quality—extending from the aesthetics of particular buildings and the overall urban form to traffic impact and the consumption of natural resources (particularly water and open space). Santa Barbara also leveraged the rules to gain affordable housing, in part by granting excess densities for developers who would provide it. In Santa Monica, resistance to growth distinctively leveraged some social service gains (in addition to other benefits), including protection for those most threatened by gentrification and redevelopment, like renters in the path of high density commercial ventures.

Santa Barbara and Santa Monica had different priorities but also different routes for gaining their particular mix of physical and social gains. Santa Barbara stressed limiting development altogether and then used those limits, sometimes unwittingly, to shape projects so as to least harm the physical environment and the supply of affordable housing. Santa Monica continued to allow high levels of development but relied on a strong left political base in addition to exactions and pioneered a variety of social benefits. Both cities participated in ad hoc exactions, but Santa Barbara had more standardized modes for handling its development linkages, whereas Santa Monica relied more thoroughly on developer agreements. Although this "flexibility" enabled these two cities to generate higher benefits from developers, the same type of flexibility could generate the opposite results in Riverside, as when a zoning change or a developer agreement (especially in unincorporated areas) led to staged projects insulated from future review.

From this broader perspective, we can see that in places like Riverside land-use policy continues to be led by developers, with minimal changes in the way things are done. In Riverside, even long-standing land-use controls, like zoning rules, were altered routinely to accommodate the requests of developers. The innovations Riverside introduced in a best case were routine in communities with stronger growth controls. Even the most direct impacts of development—on traffic and public infrastructure—were assessed in narrow terms—calculated as though the fiscal cost of each unit of additional growth would be the same as the average cost of public infrastructure and services overall. This methodology ignores the accelerating marginal costs of development as communities are required to expand overall system capacities for road networks, water supply, wastewater, sewer, and so forth. Riverside was probably not collecting enough to offset the stream of present and future public costs associated with new development—that was certainly the view of planners in the other jurisdictions. Within its immediate geographical context, Riverside was fairly advanced in its growth management techniques, but these innovations were not strong enough to change the heart of the process. This situation was probably fairly typical of the new wave of growth control localities in California, like Riverside's neighbor Corona, which had adopted enough new development regulations to exceed—by at least one counting—Santa Barbara as a "growth control" site (see Chapter 3), but where, nonetheless, growth was neither slowed nor "controlled."

Riverside's gains, however small in comparison to our other study areas, did mean some additional traffic mitigations, at least in regard to its best-case project. The city also secured protection for its citrus belt—although the integrity of this protection was threatened by city council exemptions (especially before 1987, when voters trimmed the council's ability to grant such exceptions) and by the sustained efforts of frustrated landholders who used the ambiguity of the city's general plan to legally challenge the citrus downzoning that was based on that plan. At the same time, the city approved a massive amount of development not subject to any extraordinary limitations or exactions, nor even to review under CEQA. Beyond the issue of not recouping costs, Riverside lacked the strong controls necessary to effectively generate the kind of growth it wanted; it wanted to increase commercial development and retail sales for fiscal gains, but the rate of gain did not approach the rate of residential building, which was well above robust state averages.

The cities in our study areas made gains from development through the substance (and politics) of implementation, not just through formal rules or abstract statements (Riverside County sounded good in these terms). All localities had general plans aimed at growth control, but they differed in content. Even more important, they differed in their application, and in whether they would be used as a legal basis for limiting development or, as in the Riverside case, for increasing it. Flexible structures sometimes worked to increase demands on developers, sometimes to decrease them. General plans and zoning laws were monitored carefully at times but treated as mere paperwork on other occasions. All our sites had environmental review programs; and following CEQA requirements, they all had similar overall statements of intent and policy. But again, local meaning was quite different.

In two of our study sites, and less so in the third, growth control did change the way growth occurred. Even in Riverside, particular community goals were enhanced by making development respond to the needs and desires of citizens and local government. But the ability of jurisdictions to act consistently in the face of developer power depended on the surrounding ideological and political context. Local citizen organization and power were the key to altering the terms of growth.

Notes

1. One might argue that Santa Barbara fees should be higher, since when an area is more built out it costs more to mitigate an equivalent increase in traffic. But this argument presumes that the Riverside site won't eventually be built out, in which case the project impact would be comparable in the end. It might be most equitable to assess project costs according to their share of ultimate build-out (where communities are willing to commit to such a limit).

2. Interview 034.

3. By way of general comparison, San Francisco's pioneering housing exaction in the early 1980s was set at $5.00 per square foot of commercial space (Keating, 1986). This was about half the amount the city's consultants (Recht, Hausrath, and Associates, 1984) had said was necessary to mitigate the effects of office development on housing (see Keating, 1986: 138).

4. Unless the developer was tapping public housing funds, which had their own set of income and rent requirements.

5. Instead of directly estimating the housing demand generated by a particular project, Santa Monica planners based on-site requirements on how much could be built with the $1.63 per square foot housing fee (calculated at $30,060 per unit).

6. The amount of annual subsidy per unit is 30 percent of the difference between median income and 80 percent of median income (median income x 20 percent x 30 percent, or median income x .06). For Santa Barbara—$43,100 x .06 = $2,586; in Santa Monica—$31,120 x .06 = $1,867.20. The income figures were provided by housing officials in each city, based on 1990 median incomes for a family of two.

7. A 1991 Santa Monica initiative requiring that affordability be targeted to those below 50 percent of the median income increased the amount of subsidy in that city.

8. We exclude from these totals any non-city funding provided under various affordable housing programs (including federal, state, and nonprofit initiatives).

9. Glickfeld and Levine found that "jurisdictions with more growth control measures tend to enact other measures which are designed to encourage the development of low-income housing" (1990: 40); however, they did not determine whether the existence of these growth control programs correlates directly with the actual construction of more affordable units. According to a 1989 study by the California Coalition for Rural Housing (CCRH), the city of Santa Barbara met with greater success than Santa Monica or Riverside—or, indeed, than the vast majority of localities in the state—in providing its "fair share" of affordable housing. This estimate holds even after correcting for the fact that Santa Barbara's figures are inflated by the inclusion of various housing assistance programs not included in other places' totals. Santa Barbara provided 83 percent of its fair share, compared with 50 percent for Santa Monica and 37 percent for Riverside. The figures look even better for Santa Barbara when we compare the city's "official" fair share needs with data that we collected directly on the number of affordable housing units generated by development exactions, linkages, and incentives. By this measure, the city provided almost twice its fair share of housing.

We caution, however, that the entire "fair share" metric is methodologically suspect. Under the statewide program, regions (counties and combinations of counties) determine regionwide needs and then allocate to sub-units (e.g., cities) their share of those needs. No standard procedure exists for either operation, leading to prima facie irrationalities in the data, such as the fact that small population localities like Santa Cruz County show aggregate needs approximately the same as Los Angeles County. If places like Santa Cruz end up with particularly high needs, reflecting internal political and organizational procedures rather than housing conditions, they will appear to be less successful in meeting those "needs."

10. Jencks (1994) argues that the destruction of single-room-occupancy hotels is the single largest contributor to homelessness in America.

11. Interview 015.

12. Interview 015.

13. A more complete comparison of design restrictions between the three cities would require peddling the projects through the architectural review process, where details such as landscaping, awnings, window and door treatments, exterior color, and so on are hammered out. We did not do this, but by reputation Santa Barbara certainly exercises the most detailed and demanding control.

7

Building the Rules

In 1957, local officials and growth boosters launched a campaign to promote industrial development in a 17,000-acre chunk of high desert in the Antelope Valley north of Los Angeles—not far from our Riverside communities. "It was a gala occasion. . . . The governor and the area's U.S. Representative were there, as were Miss Antelope Valley Industry and Miss Goodwill Ambassador to Industry. The day of speechmaking, terming the Valley the 'biggest industrial Empire in the West,' was topped by a sonic boom and the dropping of a bomb . . . to make the first excavation for the industrial park" (Fellmeth, 1973: 301). Today it is doubtful that local officials anywhere in California could open such an expanse of land to development without some form of environmental review and probably some extensive assurances that environmental effects were taken into account. There would surely be no desert bombing as a publicity gimmick, much less with the implied sanction of state government.

Beneath the numbers and strategies we have been describing lies one of the dramas of the contemporary United States—citizen action at the local level, repeated again and again in California and in many other parts of the country over the decades since bombs were dropped on Western deserts for grand openings. In a political landscape characterized variously as apathetic, nonparticipatory, and quiescent, this is a realm where people see things they don't like and take collective action. These actions are not just public grousing, name-calling, or other personal attacks. They take organizational life and are policy oriented. They are not "outbursts" but—in the main—considered and deliberate attempts to reorganize the way community is put together.

Quite beyond "not-in-my-backyard" opposition to a specific toxic dump, highway, apartment building, or bombing, these citizen actions can have a broad conceptual and geographic reach (Szasz, 1994: 165–166). Although not as encompassing as Marxian or Christian utopias, urban environmentalism is uncharacteristically broad by American standards.

As is evident by the range of exactions, linkages, and programs that have made them up, growth control and related urban environmental programs bring into play a wide variety of social and environmental goals.

We are not arguing that all growth control programs are optimal in either a social or environmental sense. Increasing the amount of land required for building lots without limiting demand for those lots (e.g., through controls on job-producing office employment) may only spread the sprawl a little thinner, using up more undeveloped and agricultural land. This pseudo-control is a time-worn practice all over suburban America, and in our study we observed it in Corona and other parts of Riverside County. Likewise, creating wildlife protection zones in the midst of unlimited growth in surrounding areas does not protect nature as a living system but only as an exhibit for "outdoor education," if at all. Natural systems require greater spaces to reach critical ecological masses. But even though growth control policies contain some counterproductive elements, the evidence we have gathered indicates that the net consequences of controls have been positive.

If "all politics is local" (in the phrase of former House Speaker, Tip O'Neill), and if the dynamics of growth control have been so important a part of local politics, then they are—potentially at least—at the heart of all politics. Any set of dynamics that stimulates new community organizations, produces additional public information about the social and physical environment, and enhances the participatory level of citizens deserves a careful look by those seeking clues to building movements for reform in the United States. Making the same point in even stronger terms, Daniel Kemmis suggests that "public life can *only* be reclaimed by understanding, and then practicing its connection to real, identifiable places" (1990: 6).

Enhancing the Quality of Public Decisions

Local environmentalism has increased knowledge, not only through the prodigious research that activists sometimes carry out but also through the environmental reviews that require gathering and publicizing information. Although our study sites differed significantly in the frequency and thoroughness of their reviews, as well as in the publicity given to their findings, there is no question that the amount of information available to citizens was higher in all our sites than it would have been a generation earlier. Cities with more intense growth controls had a more comprehensive and precise view of the costs of development, in part because their rulings were likely to rest on full-scale environmental review. Officials could make more informed decisions because the information shared, discussed, and debated had grown in breadth and depth.

Beyond making individual policy makers, developers, and citizens more knowledgeable of growth impacts, the changing knowledge base creates a greater capacity to solve problems collectively. This improves what Elkin calls "social intelligence" (1987; following Dewey, 1954; and Lindblom 1965, 1977). Where viewpoints are excluded, options are restricted—"problem solving is . . . likely to be ineffective simply because some desirable alternatives will go unexplored" (Elkin, 1987: 95). Rather than tallying efficiency in terms of total economic activity, one can evaluate efficiency as the capacity for informed citizenship—something that leads, says this credo, to both more sensible and more democratic decisionmaking.

In our study areas, local environmental controls brought a greater diversity of perspectives and information into the process of land regulation; not all the new input was necessarily accurate or wise, but it was unarguably broadening. In addition to ever-present desires of developers and growth boosters, decision makers are now more likely to consider the concerns of other parties in the community (local residents, environmentalists, disadvantaged social groups); these parties often care not only about the direct and immediate effects of development on the human community but also about the impacts on the natural environment itself. These shifts in concern are apparent in formal rules; more importantly, we encountered them in the reports of informants and in our observations on the implementation process.

New viewpoints provide a more complex calculus of community interest and take the place of the simplistic "trade-off" between regulation, on the one hand, and economic productivity or "efficiency" on the other. Contrary to the standard advice that has been pushed by influential academics (e.g., Peterson, 1981), it is silly to measure local efficiency by the ability of a community to attract investment. If a transaction in business is conducted at a loss, increasing the volume of those transactions brings ruin, not solvency. The same is true in the public sphere. Based on our evidence, slower-growth communities did not hurt themselves by "interfering" in the capacity of developers to make their investments where and how they wanted. Cities with tight controls still attracted investment of all types and, virtually all our evidence suggests, with far more positive effects than would otherwise have been the case. By designing purposeful barriers to entry, communities experienced better growth than what was otherwise in store. Developers sometimes call such barriers blackmail, but they could also be regarded as merely holding out for the best deal, as business people know to do.

Beyond our own efforts, studies mount, showing that in terms of unemployment, fiscal health, crime rates, and other indicators of social or economic well-being growth helps little. Indeed, there is reason to believe that fast-growing localities are more vulnerable during economic downturns.

The Inland Empire surrounding Riverside was the fastest growing and most loosely regulated of our study areas during the 1980s but was hit particularly hard by defense industry restructuring in the early 1990s. Unemployment rose to 11 percent by 1992, well above the Southern California average (Jonas, 1997: 216). This set of macro-findings further strengthens the view that localities must scrutinize the way the market comes to visit them, and that forgoing certain projects in certain places and times—"losing" them altogether—may do more to develop the local economy than pursuing growth itself.

Regulation as Economic Boon

Besides the cleverness of developers and the willingness of growth controllers to compromise, there is another class of explanation for the robust economic growth that took place in our study areas: It may be that regulation fosters certain kinds of development that otherwise would not take place, and thus even if some projects are lost to the locality completely (something we did not see, but may have missed), other projects—better ones—are gained as substitutes.

More specifically, countless commentators now urge that quality of life matters in creating economic growth (e.g., Porter, 1990). Businesses in the advanced, high value-added sectors of finance and technology—those sectors that rely the most on human capital—want to do their thing in a place with the kind of lifestyle resources that attract the kinds of employees they need—a place with high cultural capital and environmental benefits. Technical inventions in, say, computing rely on imaginative software creators, up-to-the minute industrial designers, hip marketing consultants, and advanced local markets where prototypes and early production runs can be tested. This kind of postindustrial agglomeration differs in obvious ways from a prior era's less complex reliance on a good port meeting up with a source of raw materials to produce, say, steel beams (see, for example, Duncan and Lieberson, 1970; Noyelle and Stanback, 1984; Castells, 1985). Under the new economic regimes, environmental protection—at least in some form—can promote just those types of economic development that have the best future (Power, 1992). Efforts to preserve open space or historical architecture can easily end up as economic gain, even if an alternative opportunity (like an oil refinery) is "sacrificed" in the process.

We have seen a number of cases where regulation supports specific forms of economic expansion. For instance, beyond the obvious examples of tourism and retirement, air quality regulations can lead to economic expansion. The first places to enact tougher standards for air quality emissions induce local firms in related sectors to innovate. Given broader trends of tightening emission controls spreading to other places, these same local firms have a first-mover advantage in competing in an expanding market. They may even lock in a long-term advantage if the specifications of their technology define regulatory standards and if their technology becomes the basis for the manufacturing of related equipment (e.g., the pipes that lock into a filtration mechanism). As with the creation of the now standard "QWERTY" typewriter keyboard, any competitive way of doing things must contend with established convention and with the inertia that comes when diverse producers, users, inspectors, governments, and firms find a solution that satisfies everyone mutually.[1]

Santa Barbara's effort to control oil projects provides some evidence for how regulation creates new technologies and business opportunities. Some of Santa Barbara County's strongest controls were developed specifically with the offshore oil industry in mind, particularly the large-scale support facilities (e.g., processing plants) built on land. Tight inspection requirements gave rise, for example, to new local technologies for underwater inspection of pipelines, underwater communication devices to service the inspectors, and a new type of underwater helmet design. The local firm making the helmets now has the standard design for all underwater uses and has emerged as a significant global company after beginning as a two-person operation at the founders' home garage. The underwater radio company now is a global operator in communications, with little of its business related anymore either to oil or to local development. The technologies created for underwater scanning are now applied to all manner of submerged pipelines, including municipal outfalls. Although a search for such innovations and their business careers was not a subject of this research (but see Molotch, Woolley, and Jori, 1998), we suspect that other examples like this would be found in the Santa Barbara study area, in Santa Monica, and in other places that drive hard development bargains.[2]

Grounding Sustainability

Concrete local environmental policies produced the practical gains that most affected outcomes across our study areas. These gains were not based on some ineffable vision of love for Mother Earth but on a simple and pragmatic approach to regulating development. If we were to provide one succinct summary statement to characterize these measures it would be that they were efforts to price projects at their real value, setting prices to reflect spillover or neighborhood externalities as well as the informed eye could see (Navarro and Carson, 1991).[3] As environmental leader Dennis Hayes put it, "One of the clearest lessons of the last quarter century of environmentalism . . . is that if you get the price wrong, any other cure is only second best" (Hayes, 1995: 1). We have no quarrel at all with the congressional Republicans' 1996 effort to require economic impacts (benefit-cost analyses) of any federal pollution rules so that regulations can be set to preserve economic vitality. But our guess is that if such accounting were done properly—and included putting realistic prices on public assets lost to future generations—developer exactions would increase vastly, rather than decrease. Even within the most narrow economistic reasoning, the expectation that development should "pay its own way" improves rather than distorts the market as an information system. The vast majority of regulations that we have observed across our study sites, and that we suspect make up the environmental regulatory regime more generally, are of this sort.

The impacts of growth control go beyond local economies, of course. When localities take action to decrease water use, for example, they not only save groundwater supplies and protect local wildlife but also avoid importing water from faraway places whose ecology can be spared. Such was the effect when Santa Barbara County refrained from participating in the California state water project. When a Santa Monica developer promises that, if his hotel is approved, he will allow no use of Styrofoam, the earth will experience that much less of the material. By rewarding such offers, the city facilitates more ecologically minded approaches to doing business—quite beyond the Styrofoam issue itself. More generally, the political mood in localities with growth control encourages various initiatives having to do with energy conservation, waste management, and pesticide use that lead not only to benign local policies but enhance the earth in more basic ways.

Beyond being bandied about as a moral principle, sustainability must be incorporated into everyday experience and the built environment in order to matter. The local construction of buildings and the configuration of settlements creates cities and societies that do or do not take more from the earth than the earth can regenerate. If, as Daniel Chiras (1995: 205) argues, "sustainability is not just another environmental issue, [but] *the* issue of our times," then local practices are a crucial locus of reform. There is at least a chance that people may extend their concerns from their own backyard toward the needs and problems of community, region, and nation—with different individuals reaching farther, according to their own sensibilities and to the nature of the issues that first brought them into the public realm. Rather than being inspired by national campaigns or mobilized by the "big ten" national environmental organizations, people start where they are at and work outward to larger issues and higher scales of governance. National environmental leaders such as Louise Hicks, Lois Gibbs, and Richard More have come up from the grassroots and have emerged from local concerns over the specific conditions of life (Almeida, 1994; Aronson, 1993; Krauss, 1989; Szasz, 1994: 89–99). And even if people who get involved initially at the local level do not go on to become national leaders themselves, they often depend on such leaders for aid in even minor skirmishes and thus come into contact with people who have a larger perspective.

Who Wins What?

Growth Control and Social Justice

What do the kinds of controls we have evaluated imply for social justice, for extending the benefits of life to those traditionally left out? However limited environmentalism and, more specifically, growth control may be, it is wrong to think of them as detrimental to the lives of low-income people. Long before the modern growth-control movements, low-income people suffered the worst environments, with telling effects on their health and well-being. The vast slums of the great cities were an effluent of growth rather than of environmental controls. The record shows that unplanned development yields vast degradation. And the United States, a rich country built under weak controls, has long been a country whose cities are the most remarkable for their high levels of segregation, privatization, and violence. In contrast, the folks "on the hill" have always known the most about what their factories spew into nature and have had the means, through spatial location and the capacity to litigate, to avoid the worst excesses of industrial activity, as well as to secure themselves from the social effects of inequality.

What is new is that with growing democratization of land regulation, marginalized communities have gained tools for resisting effectively the sitings of undesirable land uses in their midst. Rather than dividing the rich from the poor and whites from people of color, environmental regulation follows from common interests.

During our study period, low-income communities and environmentalists in both Santa Monica and Santa Barbara tended to vote together in local elections; both sides assumed that environmentalists gained their strongest majorities in working-class precincts and in those with the strongest participation of African-Americans and Latinos (albeit with low rates of voter turnout, which environmentalist campaigners tried to raise). In some other cities, the formal leaders of poor and minority neighborhoods have apparently allied themselves with growth interests (Stone, 1989), but such coalitions were not evident in our study sites (or, indeed, in others; see, for example, DeLeon, 1992a, 1992b). Instead, symbolic efforts by business groups to bring in unions and minorities had little success, at least at the ballot box. True, most environmental organizations in our study areas were led overwhelmingly by middle- and upper-middle-class white people—with high numbers of women serving as volunteers and providing a great deal of the money and the skills. But this middle-class profile parallels the leadership of virtually all voluntary associations: These are the people with the time and resources to participate (Milbrath, 1965; Verba and Nie, 1972; Warren, 1963). Middle-class organizations should consciously do more to reflect and represent the underlying coalitions that form their base of support. At the same time, leaders of more traditional urban constituencies, such as minority and working-class neighborhoods, would be wise to link revitalization and equity initiatives to growth issues in ways that parallel environmental justice movements (Warner, 1997).

By enacting limits on development, localities are genuinely in a position to choose the kind of development that will do their citizens the most good—whether in terms of jobs or housing or recreation or some other valued outcome. Both Santa Barbara and Santa Monica placed high priorities on generating affordable housing. Santa Barbara's control over the development process provided a major source of such housing; in one example, developers were enticed to include affordable housing by the prospect of doubling the allowed densities (densities, in this case, that were higher than would have been allowed in Riverside). Such incentives hinged on already established limits to growth, without which—it is safe to say—much less new affordable housing would have been built. Indeed, California data indicate that the cities of Santa Barbara, Santa Monica, and Riverside provided more than their "fair share" of low-income housing (California Coalition for Rural Housing, 1989). Perhaps this provision of low-income housing is the result of taking planning more seriously: Strong planning correlates with strong concern for meeting local housing needs. Or perhaps, as we have implied, strong development controls gave city planners the leverage needed to gain the housing.

Our study sites do not exhaust the variety of community agendas that might orient local land development regulation. For example, growth control in Santa Fe, New Mexico, came to emphasize the preservation of Chicano community and culture against a tide of artsy investors, new-agers, telecommuters, and retirees (Tatum, 1995).

Who Can Harness Growth?

The successes of Santa Barbara and Santa Monica (compared to less evident gains in the Riverside study area) are not utterly random—folks in any old place are not as likely as any other to have such achievements. The stereotype has it that rich people are key. But it is not clear that wealth drove strict growth control in our study areas. Santa Monica was a highly diverse city, and even the mean income in posh Santa Barbara was below the state mean—although the story here is complex, since, as we indicated earlier, a significant number of extremely wealthy persons did play an important financial role in the city's development.

Still, even in the case of Santa Barbara, it is not just wealth that explains the successes of growth control but also a robust tradition of activism and civic organization—it is in part a land-use civic tradition that directed Santa Barbara's historic spin. And, later, Santa Barbara became a university city, which not only affected the voting base but spawned educated activists from student, faculty, staff, and spousal ranks (in this, events in Santa Barbara ran parallel to events in places like growth-control oriented Boulder, Colorado). Then there was the environment-oriented tourist economy, which spun its own allies, however strategic and narrow in focus.

Santa Monica—with its weaker civic tradition, its lack of a super wealthy class, the absence of a major university, and its interconnectedness with Los Angeles—was the less obvious site for reform. And because a majority of residents rented rather than owned their homes, Santa Monica was also deprived of the kind of large-scale home ownership that is ordinarily associated with stringent growth controls (Protash and Baldassare, 1983). Yet Santa Monica, in part via the tenants' rights movement, did adopt significant growth control measures, which highlights the likelihood that more than one kind of place profile can yield land-use reform.

In spite of the differences between Santa Barbara and Santa Monica, the two cities do have a commonality that we would like to advance as something that should be considered in large-scaled analyses of growth-control across localities. Santa Barbara and Santa Monica were localities in demand. Investors very much wanted to develop land and to build buildings within the boundaries of these cities. Santa Monica became a center of commercial expansion and, increasingly, a place where higher-ups in the service and entertainment industries wished to live. As with Santa Barbara, this demand both fueled rapid growth and roused citizens to think they might be "overrun" if they did not begin to guide development, if not curtail it altogether.

In this case, the common denominator was not the income of citizens but the comparative advantages of these places over others as places to live and as places to own a piece of. Given the presence of investors who wanted to come into these cities, those who regulated development could pick and choose, or at least redirect projects in desired directions. Again, we are not proposing that demand is a simple explanation; many places with this profile have not resisted growth at all—that's mostly why urban America has the form it now does. But it seems that investment demand, in the recent era, has been a common characteristic of places that have been led to make the most noticeable moves toward shaping their growth (Boulder, again, is a consistent example).

Comparative advantage is one of the ways that local changes may be "located" and understood within broader contexts of economic restructuring and globalization. At the same time, we should examine how other historical conditions figure into transforming the local conditions of growth. Ferman (1996) argues that local institutional and cultural factors distinguished challenges to the growth machine in Pittsburgh and Chicago. Walton (1992a: 276) shows how local environmental successes in the Owens Valley of California were built not only on local traditions of resistance but on the culture and institutional structures of national environmentalism, specifically NEPA and CEQA.

Giving Other Places Leverage

What about places lower on the investment-site scale—places where people of lower income do unglamorous work and where, most crucially, investors have only lukewarm interest? Are these places not left out? As privileged localities gain more amenities, don't they distance themselves from poor localities? Won't this "stratification of places" (Logan, 1978) just grow more intense, creating even more unequal life chances among Americans—based on where they happen to live rather than what they do for a living?

Actually, we found that the emergence of growth control created certain positive spillovers for less advantaged localities, a point sometimes noted by planners and environmentalists in such places. In any ordinary competition between localities, the advantaged communities can afford incentives to draw investment their way, leaving their neighbors with less desirable uses. But in our study areas, we found that under growth restrictions advantaged places did not merely take everything; their choosiness worked to their less well-fixed neighbors' advantage. First, the front-runners instigated some controls that affected all localities: For example, the requirements that all cities and counties have general plans, that there be environmental impact reports, coastal protections, and controls over air emissions and toxic disposal all tended to emerge from the more environmentally oriented localities, places whose representatives in Congress and in the legislature pushed for general protections as well as protections specific to their constituencies. To pick a more specific example, the California coastal initiative, passed at the polls by the largest majorities in the environmental bastions, applied to the whole state, including its disadvantaged seaside communities—of which there were a good number (e.g., Eureka in the north, Long Beach in Southern California, and Avila Beach in the state's mid-coast).

Environmentalists in places like Santa Barbara were not only active on the local scene but also played key roles in national environmental organizations like the Audubon Society and the Sierra Club (one of the local activists, for example, had been conservation chair of the national Sierra Club under David Brower); these national organizations pressed for statewide and national rules. Congressional representatives and those elected to the state legislature from such places tended to push for environmental protections, even if they otherwise took politically conservative stands. In order to gain their constituents the protection they demanded, the politicians ended up championing legislation that had much wider impacts. For example, the primary rules affecting offshore drilling in the United States were a response to Santa Barbara's demands, which, in turn, were pushed by their Democratic Party representatives in the state government and by their highly conservative Republican representative (Robert Lagomarsino) in the House of Representatives.

These privileged localities also served as litigation "test sites." They were the first to face court challenges from development interests. By financing and fighting these legal battles, these localities refined rules and legal theories and made them increasingly bulletproof. With the legal precedents already set, such measures could then be taken "off the shelf" by other jurisdictions.

Over the longer term, less advantaged groups may also benefit from environmental control, as neighborhood succession brings them better places than they would otherwise inherit. Historically, the poor and minorities have been relegated to those areas that were outdated, if not defiled. Industries have dumped toxins into their soils; governments have built freeways near their lungs; their building walls have had the most layers of lead paint. However, elevated standards of environmental design produce places that are of higher quality for those who have the lowest buying power. It is over the long term that many environmental controls have their most important benefits. Indeed, sometimes the difference between strict rules and weaker ones is the time frame, and the unanticipated future use of a given land parcel or structure. Scrutinizing projects in what are today's affluent areas may well protect against the risks and the negative spillovers that would come to light in the future, and especially in the lives of the poor (for discussion of related issues, see Piller, 1991: 158–172).

Finally, in organizational behavior generally and, we presume, in environmental policy making specifically (Dimaggio and Powell, 1983; Scott, 1987; Tolbert and Zucker, 1983), the simple fact of imitation is an important element in policy innovation. Once a number of localities establish new rules, especially places regarded as "leading," other localities follow suit. Planning professionals in the less advanced zones can point to what other places have done. More than fostering mindless imitation, these precedents serve to up the ante for what is considered "decent" or "reasonable." Many places can then get on the bandwagon, as the norm shifts toward preserving local community and environmental qualities. One of the reasons these new norms may actually take hold is that when developers have been turned down in one place, their bargaining position is weakened vis-à-vis their second-choice locality. In this way, if localities at the top of the desirability scale stand tough, other places are in a position to get more than usual from developers. As with other dynamics we have discussed—technological innovations, organizational acumen—this rebalancing of bargaining power operates at the global scale as well. Anything that curtails the site options of polluters may help spare poor people some of the horrors of runaway shops that poison. To return to an earlier theme, action at the local level—in this case in the more privileged places—has impacts on the global system, even in localities at the opposite end of the vulnerability spectrum.

Cutting the Costs of Interlocal Competition

Without toughening regulations at the top, every place is in a competition for limited amounts of development—a dynamic that, despite the loud worries about growth controls, is still far more evident and robust than the opposite tendency of places blocking development. In the competition to attract projects, places with lower comparative advantage are also limited in how much they can really afford to give back, in their in-house capacity to evaluate true costs and benefits, and in their ability to withstand the political pressures to settle for bad bargains. They are most likely to do themselves in. Within our study areas, the real payoff from these deals was an open question, as it is elsewhere. Santa Barbara and Santa Monica offered "incentives" to developers, and even though they could choose to do so for only those the projects they thought would be most desirable (e.g., tax-rich retailing or high-wage office functions), they had only approximate knowledge of benefits. The subsidies they offered were probably due as much to the drumroll of businesses' demands to "do something about the economy" as to a dispassionate economic analysis of project consequences. In other words, these cities were able to avoid the worst kind of bargains that play out in other places, but that doesn't mean they made optimum deals.

For ordinary places, investment chasing comes at much greater risk. Commenting on the United Kingdom, Savage and Warde (1993: 172) say:

> At present local authorities dare not cease to advertise their attractions, cajole mobile capital and provide limited incentives, largely because other places are doing these things. The result is probably expensive, essentially futile, competition, as the location decisions of capital are not much influenced

(most accounts of regional economic policy suggest this) and anyway success mostly means taking investment that would have gone somewhere else in the [country].

Missing from the ordinary calculus is a consideration that should be central to regional, state, and national planning policies: the net effect of interlocal competition on overall well-being. If San Antonio gains a wonderful convention center, some other city—maybe one that has already built its convention center and must pay off its bonds—will gain fewer conventions. The danger at the broader level is an irrational overbuilding of certain kinds of infrastructure, an oversubsidy of growth projects, and a ratcheting down of what the general public gains from the country's total development activity. Especially in an anti-tax period, about the only way to support public services is to get that support out of projects "up front"; if that moment slips by, the chance may well be forever lost. Give-backs to developers, in the form of subsidies or a relaxation of standards, only exacerbate an already losing situation. Under such conditions, large volumes of development create only large volumes of long-term loss.

This problem can be solved in the arena of national policy. The U.S. government could stop providing grants for projects that feed interlocal competition. More ambitiously, the federal government could eliminate tax exemption for bonds that help finance these schemes. More radically, the national government could, even if it took a constitutional amendment, forbid the use of special deals to support poaching across jurisdiction borders.

Sad to say, the substitution of a Democratic president (Clinton) for a Republican one (Bush) did little at the federal level to alter this way of doing business. Rather than taking an interest in the overall economic and environmental well-being of the country, urban policy—such as it has been—has championed still more private-public partnerships in which localities are, at best, merely trading off losses and benefits with one another. The celebrated secretary of the Department of Housing and Urban Development (HUD), Henry Cisneros, was selected in the first place for his success in "building" San Antonio, Texas, in just this manner during his tenure as mayor. Cisneros urged all cities to be like San Antonio, disregarding the zero sum nature of this competition between places and the tendency to lower standards on all urban fronts. Together, Clinton and Cisneros maintained the casualty-strewn urban war of all against all.

Making Markets Smarter

Greater local political involvement and better knowledge of the stakes of the development process inhibit growth giveaways and also hold out hope for the larger community of national interests. Localities gain not so much from a particular rulė or zoning change or utility control but from a general politics that is skeptical toward development. Developers complain bitterly, but the "wall of no" pushes them toward the kinds of projects that satisfy local preference—and hence toward a different kind of "yes." Developers bear the cost of often vague and changing standards, to be sure, but all this is a sort of proxy for market uncertainties. Rather than localities competing to sell themselves to developers, the developers are competing for public favor. The "customers" whose preferences must be fulfilled are not just the individual end users (such as home or business owners) but the broader community as it seeks to manage and minimize damaging spillover. The community or neighborhood or local government, in effect, shops for the most appropriate projects. The developer's job is to cater to the end users but also to all the "neighbors" that will endure the collective costs and benefits. The job of selling a project becomes more complex and demanding, but nobody ever said it was going to be easy. It is a way to think about how markets and regulation can work in tandem. In creating a private development, the entrepreneur must sell it twice: first as a public good, and then to the individual end user as a private good.

The free marketeers underestimate the market. Markets can handle regulation, at least when citizens accept the basic legitimacy of the state and of the markets it regulates. On the other hand, corruption can permeate both spheres, with government corrupting the market as the market corrupts the government. Contemporary Russia, China, and the Italian building industry (Molotch and Vicari, 1989), to select some well-known examples, have had such a corruption symbiosis. When urban regimes are honest—and we believe that in most localities they are—government can enhance rather than corrupt market efficiency. The market, in turn, can respond to rather than corrupt public goals.

Our studies of the practice of growth control reveal how powerful interests adjusted to a new, emerging institutional framework. The shift in power created a different "envelope" within which the market worked, an envelope that included various political, legal, and financial resources that were available to the growth advocates as well as to the environmentalists. The developers' lawsuits and legal strategies were just as much a part of the regulatory reality as the laws they challenged, so were their omnipresent campaign contributions, sharply increasing over time, as well as the pro-growth editorializing of the daily newspapers in every town and city we investigated, including the most anti-growth locales. Development interests did not lose all their power, by any means, as evidenced by the robust growth we observed in all our areas. The on-the-ground result shows that rather than stopping or, in most cases, even curtailing development, environmental reform—when it had any effects—incorporated new environmental and social considerations into the growth that took place. In other words, the market persists as a crucial institution, but now within a different sort of local framework. The market's robustness is demonstrated by the way entrepreneurs suck up opportunities left open, including a number of opportunities unanticipated by the regulators. More profoundly, the power of the market shows itself in the way regulation converts the city into a customer that entrepreneurs seek to satisfy. They do satisfy this customer; projects are built; and builders make money.

Valuing Land and Community

Clearly, the analytic division between use and exchange values does not equal a simple boundary between corresponding political interests on the local scene. Political motivations and power-building are much more complex and contradictory than that. The path for land development and growth has worn a clear groove through the institutional practices that shape land use and planning in the United States. Even under the tightest sets of building rules, developers make the most of a continuing systemic bias that favors growth—as described in Chapter 4 and as posited by growth machine theory and urban regime theory. Developers also tap structural resources such as money, access to media, and organizational capacity. As we have also seen, the terms of growth can be shifted, bringing new stakeholders into the regulation process and producing a range of distinct environmental, social, and economic results for communities. Building does not rule by fiat but continues to be a central dynamic for communities, shaping local power and the built environment.

In the political push and pull of planning for communities and dealing with daily decisions on project proposals, new combinations of exchange and use values emerge—"enlightened" developers work with environmentalists and affordable housing advocates, builders deliver anti-government and anti-elitist appeals to homeowners and labor unions. Against this backdrop, policy makers and planners are not simply agents of growth interests, but also people who advance their visions of good planning within their definition of political and institutional limits. Values of land are being combined in different ways, but the distinction and the tension between use and exchange orientations remain embedded deeply in mainstream culture, in institutional practices in the United States, and, in different ways, across the modernized world. As long as this is true, growth machine theory will be a powerful analytic tool (see also Molotch, forthcoming).

Local politics is the arena for this transformed relationship between state and market. This claim runs against arguments that local politics have only declined in significance, as evidenced by dependence on national and global economic forces and as marked by declines in voting, by weakening of party machinery, and by pullbacks in social provisions. When localities step out of the limits that go with chasing investment, they can be a meaningful force as a customer and proactive in influencing the very market that otherwise dominates them. Indeed, under the U.S. system of land-use home rule, they have the most political and legal leeway to set the terms of investment—made even more necessary by state and federal cutbacks, which have removed the subsidies that formerly offset some of the costs of local development.[4]

Treating the community as customer may seem an analytic sleight of hand. Ordinarily, only individuals (including corporations now conceived of as akin to a person) are thought of as consumers. In the hard-headed world of markets, "community" usually takes on a soft, wishy-washy status, one that developers invoke only when they appeal to civic pride to build support for their projects. When environmentalists resist on grounds that the costs of the new uses will hurt, and that marginal costs should be borne by those who will benefit, they are accused of being "selfish," of "pulling up the drawbridge." In effect, developers invoke community to support the idea that everyone should share the costs of growth. They hold implicitly a model of a three-way communal bond among those already in place, those sponsoring the new projects, and new users who will be served. Ideally, from developers' perspective, community is a mutual support system.

Otherwise from these quarters, community has meaning only as a "selling feature," rather than as an animated, active force in the world that has its own interests: a shared set of sensibilities that can, through the political process, have a policy expression. In Paul Peterson's (1981) sympathetic academic formulation (following Tiebout, 1956), individuals shop for place the way they shop for trinkets and laundry soap. The place that most satisfies popular taste will attract them the best; hence laissez-faire in land-use matters is probably the best policy for optimum land-use results. Although it would take us far beyond the focus of this book to argue the point fully (but see Logan and Molotch, 1987), we conceive of community as something in which people have deep personal and practical investments. By virtue of forethought or accident or some combination of the two, residents are not "free" to substitute one sort of city for another; they are bound to the place where they are or to the place they must move by complex, overlapping, and intersecting needs of love and work. The likening of community to ordinary commodities that can be selected or discarded by simple preference goes against all we know about the complexity of people's daily struggles and about the way their instrumental routines, affiliations with one another, and affections toward their physical environment provide meaning to life.

Balancing Commercial and Public Interests

Taken singly or as a group, developers' do not usually make it their business to understand much less enhance the collective life of urban citizens. Developers who embrace this sort of role, whether in pursuit of neotraditional neighborhoods or some other valued community, are a rare exception. Indeed, the building industry does not do so well at protecting its own "selfish" interests (they often soil their collective nests). If not Leviathan, then some other mechanism must come forth to coordinate for the commonwealth—even of the developers themselves. To take the most mundane issues, government uses taxes to create an infrastructure of roads and utilities that make real estate value possible. Zoning keeps the foulest of uses from ruining market values for everyone else. According to Elkin (1987), the founders of the American republic gave commercial interests a special stake, but they did so not simply to further their own class interests. They thought the business sector could provide the prosperity to sustain a democratic citizenry, but of equal importance, commercial interests could counterbalance ideological "tyrannies of the majority." The role of government soon became one "not just of securing property but of *promoting* a commercial society" (Elkin, 1987: 116). But promoting commerce is not the same as catering to business interests: "The founders . . . thought that businessmen's views were likely to be narrow and self-interested when it came to fundamental questions such as how economic growth should be promoted and how wealth and property in society ought to be distributed" (Elkin, 1987: 118). For the system to work, citizens need to interpret and advocate *public* interests effectively in order to counteract concentrated business power. Gaining and using this capacity is the formative dimension of politics that mattered to such classic political theorists as Tocqueville (1945) and Mill (1974) (for a contemporary version, see Kemmis, 1990: 119) but that is so often forgotten today. In debating local land-use policies, citizens grapple with the commercial and public interests in quite tangible terms, in part because they have to live with the results.

Political and civic leaders appear increasingly ready to sacrifice local communities and natural environments on the altar of economic productivity and competitiveness. Business leaders sound the siren "wake-up call" for deregulation and privatization. They ask that "governments be run like a business," by which they usually mean that governments should compete against one another for the sake of *their* business. But running government "like a business" would really require that governments get the most for their "share holders," the citizens. It would mean, as we have suggested, holding out for maximum advantages in the development process, rather than minimizing them. It would mean forgoing transactions that would fail to show a net public gain. The regulatory regimes that we have described signify a generation of intervention into the urban and natural environments and show the value of just this approach. This era of land-use intervention has worked—and should be remembered as having worked—not because it has destroyed markets but because it has channeled them by making the community itself into customer. The main difficulty has not been that interventions have gone too far but that they have been applied so unevenly across localities. Growth controls—and by extension, other forms of environmental reform—hold out new possibilities for fine-tuning markets and bettering life, especially if the lesson is learned and applied more widely.

Taking Stock of Local Controls

Over the span of the years we studied, this "golden age" of environmental regulation, the image of local rule-making shifted. Regulation became a scapegoat for other problems in the political economy. Although the dynamics of global and national productivity are beyond the scope of this book, we can say that in the case of California the state's especially steep job declines during the recession of the early 1990s have been quite simply explained. Notwithstanding the claims of business and of the politicians, losses in defense contracts, a worldwide decline in aircraft orders, and falls in construction (based, in turn, on the defense-aircraft sector drop) account for virtually all the difference between California and the rest of the country during this period (for compelling documentation, see Levy, 1994: 3–14, 3–16, passim). California's sharp economic decline was not likely a result of conspiracy—environmentalists pulling off a land-use coup—but more a consequence of the complex political and economic forces that brought an end to the cold war.

The thrust of our argument has been to deny that growth control has stopped economies or even done much to slow them down. However, we have also argued that these controls have had the real—and, in our view, beneficial—effect of moving development that does take place in directions that enhance both the environment and social equity. These are substantive issues that can take us beyond arid debate to the fertile grounds of community building.

We are not saying that these controls have been revolutionary or that they have solved the relevant problems even where they have been applied most completely. Santa Monica's use of development to aid the homeless, an ad hoc remedy dependent on the right projects coming along and on a tolerance of panhandling, has not eliminated homelessness, nor the underlying problems of poverty, substance abuse, and mental disturbance that put needy people on the streets. Despite its ambitious housing focus, Santa Barbara still has affordability problems, creating a substantial number of homeless who simply lack the wherewithal to buy shelter (Rosenthal, 1994). We also doubt that Santa Barbara's rigorous exactions have improved the local environment, compared to the options of no projects at all or the widespread adoption of greener technologies. Nor do we presume that these projects pay their own way in fiscal terms. But we do think that these modes of development are surely improvements over what otherwise might take place and that they point to the kinds of solutions that might work on a larger scale and over a longer time frame.

We are arguing that the *direction* of impacts was positive, not that all problems were solved. We make this case for the record because the way these programs are remembered, the manner in which this regulatory moment is recalled, will likely have an enduring impact on future possibilities. Economic downturns have a specific discourse that can set political and economic trends for years to come. Americans who remember the Great Depression envision unfeeling plutocrats who miscalculated economic forces, were incompetent at altering the nation's course, and were insensitive to good people's suffering. Something like this orientation gave not only Franklin Roosevelt his first term as president but also the United States whatever it was to have in the way of a welfare state over the next fifty years.

If public memory of post-1960s land-use regulation is that jobs were lost, economies were destroyed, and the poor were done in, then there can be almost no hope of taking advantage of the actual experience. But if these programs are remembered as useful works in progress, they can point to new ways to develop—in a full sense of the term—political economies of place that incorporate nature as well as the quality of life into their social intelligence. The largest impact of these growth control programs, therefore, might not be the wetlands saved, nor the housing built, but the realization that it is worth it to try to pursue agendas of public benefit through the development process.

Notes

1. This use of the idea of "inertia" comes from Becker, 1995; Latour (1987) discusses the "lashing" together of complexly differentiated activities; organizational propensities to "satisfice" rather than optimize is a key idea in March and Olsen, 1976.

2. Environmental businesses have reached a point where they can constitute their own industrial districts. For example, the environmental technology complex of North-Rhine Westphalia in Germany is said to have grown to more than 600 firms accounting for around 100,000 jobs (Grabher, 1990, as cited in Lash and Urry, 1994: 92).

3. Navarro and Carson (1991) argue that a full range of environmental costs and benefits, rather than the narrow studies of housing prices, should be considered in evaluating growth-control reforms that have dominated policy research. They offer some "guesstimates" of the tremendous impact of environmental consequences when translated into dollars and cents and suggest techniques of measurement and quantification for further research.

4. As our colleague Richard Box pointed out, these circumstances set Paul Peterson's (1981) ideas of "city limits" on their head, with localities having as much or even more room to innovate as state or federal officials.

Appendix A:
Measuring Growth Control Impacts

Many people have assumed that growth controls have direct impacts on the amount and the pace of growth. After all, this was the explicit intent of some of the new regulations. Those who worry most about the intrusion of new rules into the "free market" have extended this logic to the range of growth management practices. Academic researchers, particularly economists, have devoted much attention to this question—with mixed results, as described in Chapter 3. To assess the consequences of growth controls, we developed an approach that considers the impact of a range of measures within our study areas. In this appendix we lay out this method in some detail and present our complete results.

The dependent variable is the yearly level of residential construction. This variable is measured by considering the number of new housing permits issued in a locality in a year. To compare this number across localities, we create a ratio of the number of new permits as a fraction of the total existing housing stock in each place. A value of 0 would indicate no new residential development; a value of 100 would signify a doubling of the local housing stock in one year.

We compiled data on eleven jurisdictions; nine were municipalities that existed throughout the twenty-year study period, and two were suburban areas within unincorporated county land for which continuous data was also available—Goleta in Santa Barbara County and the unincorporated parts of western Riverside County.[1] Rather than trying to specify every factor that might determine building activity, we held these other factors constant with several broad variables to isolate the impacts of growth controls. We modeled a regression curve for our cross-section of jurisdictions from 1971–1990 and tested the significance of each growth control compared to the effects of statewide trends and other local characteristics.

The primary independent variables in our analysis are the new growth controls that might impact residential building activity. Our interviews and related documents yielded fifteen distinct policies thought to control or manage growth (primarily residential, as it turned out). Drawn from four cities (Corona, Santa Barbara, Santa Monica, and Riverside) and one unincorporated county area (Goleta in Santa Barbara County), these controls were high profile within their respective jurisdictions. The fifteen policies included moratoria, population limits, downzonings, and infrastructure requirements (see Figure A.1).

The control variables in our analysis were the statewide building rate and the other unique characteristics of each place that were captured in a dummy variable

FIGURE A.1 Growth Control Measures

Generic Growth Control

This variable encompasses all of the below measures. It is scored as 1 if any of these specific controls are in effect.

City of Santa Barbara

Santa Barbara 1 (1973) Downzoning of residential areas—city council reduces the maximum residential build-out in the city by lowering the allowable building densities in the residential zones. This was passed first as an interim measure but was later ratified by popular vote in 1975.

Santa Barbara 2 (1977) Population goal—voters adopt a population goal of 85,000 for the city and at the same time require that amendments to the general plan be approved by popular vote.

Santa Barbara 3 (1986) Interim water controls—new residential permits were restricted to fifty per year, allocated by lottery. No new water use was allowed for commercial development.

Goleta (Santa Barbara County)

Goleta 1 (1972) Goleta water moratorium—the water board ruled that no new water hookups should be granted. This measure was later supported by ballot initiative.

Goleta 2 (1978) Population goal—county voters passed an advisory measure for population growth to be held at .9% per year.

Goleta 3 (1986) Goleta residential limits—county supervisors pass a resolution for an overlay zone in Goleta, holding residential growth to 1.2% per year (roughly 200 units). This later became the basis of the Goleta growth management plan in 1989.

The City of Riverside

Riverside 1 (1979) Proposition R—creates citrus belts and hillsides zones allowing only very low-density residential construction.

Riverside 2 (1982) Sewer limits—the city limits the number of new residential hookups to 1,200 to 1,500 per year (later converted to 2.5% per year).

Riverside 3 (1987) Measure C—voters reduce city council's ability to make exemptions to the citrus and hillside protections (requiring a supermajority council vote on such decisions) and require the city to manage growth in annexed lands.

The City of Corona

Corona 1 (1976–1977) Building permit moratorium—the city caps residential building permits to n per year for two years, because of inadequate sewer capacity infrastructure limitations. (This variable was not lagged since it directly effects the number of building permits issued.)

Corona 2 (1979) Permit point system—city council introduces a point system for granting building permits to encourage contiguous rather than leapfrog development.

Corona 3 (1986) Minimum lot size initiative—voters increase the minimum size for a residential building lot in the city by initiative.

The City of Santa Monica

Santa Monica 1 (1981) Building moratorium—city council passes residential and commercial building moratorium then requires all larger projects to be approved by the planning commission regardless of zoning.

Santa Monica 2 (1983) Housing element—city adopts a new housing element, including a requirement that 25% of all residential developments be affordable.

Santa Monica 3 (1987) Residential moratoria and downzonings—the city passes a permit moratorium for the Ocean Park, pending a downzoning. This is followed in 1989 by similar action in the Willshire to Montana neighborhood.

FIGURE A.2 Ideal Model

Local Building Activity = Statewide Building Activity + Local Characteristics
+ Local Growth Controls + Unexplained Variation

---

for each locality. By including a dummy variable for each jurisdiction, we controlled the local variation in building rates that was unrelated to specific controls. These "place variables" capture the range of local characteristics (social, economic, political, spatial, and cultural) that may impinge on building rates independently of the enactment of specific growth controls. We controlled for broader economic factors by considering the rate of residential building activity across the state as a possible determinant of local growth and as a proxy for pertinent statewide economic trends (e.g., business cycles, interest rates, tax law changes). The state building rate was measured in the same way as the local residential building rate—residential units permitted/existing and previously approved housing.

We evaluated policy consequences by comparing the level of residential building in situations where policies were and were not in effect. Dummy variables for each growth control assumed a value of 1 for the locality affected for each year the policy was in effect and 0 in all other cases. We needed also to consider the delay or lag time before enacted policies really take hold; impacts are often not felt until some time after they are first instituted. This degree of lag may vary from place to place and among regulations. To cover the reasonable possibilities, we carried out analyses using one-, two-, and three-year lag times from date of enactment.[2] Because of the similarity of results, and to simplify exposition, we report detailed analysis based only on the two-year time lag,[3] noting significant differences using the other lags when they are relevant.

Statistical Techniques

To avoid some of the methodological weaknesses of earlier research, we collected data that are both cross-sectional and longitudinal (technically, pooled, cross-sectional data). This sort of data, however, requires more complex techniques of statistical analysis. Ideally, we would have liked to consider a single model that tested the effects of each of the fifteen growth controls on residential construction, while controlling for the unique characteristics of each place and holding the state building rate constant (see Figure A.2).

Pooled cross-sectional data, however, often violate certain assumptions of the standard Ordinary Least Squares (OLS) regression models, as we confirmed in this case through diagnostic testing. With pooled cross-sectional data one must consider possible violations of the standard OLS regression model, as the error terms may be serially correlated, heteroskedastic, or both, as they were in this case. Serial correlation occurs when some unmeasured factor (say the volatility of local development approval practices) links what happens one year with what happened the previous year (in either a positive or inverse fashion). Heteroskedasticity occurs when the variance of the error terms differ across the cross-section of localities. For example, some jurisdictions (measured by the place dummy variable) may have much greater variation in building activity than others (for discussion of these issues, see Pindyck and Rubinfeld, 1981). These violations of OLS assumptions flaw the significance tests, as we discovered in this instance through sensitivity analysis.

We therefore used the SHAZAM computer program (K. White, 1988) to produce regression estimates suited to this type of data, using a General Least Squares (GLS) model as described by Kmenta (1986). The estimated effects of each potential determinant of building activity and tests of statistical significance (coefficients and t-ratios) can be interpreted in the same way as OLS regression statistics. To carry out this more complex, but more appropriate, statistical analysis we were not able to include all of the variables in one model. Instead, we ran three separate regressions, including a simplified model and two more complex variations. We considered the results of each in reaching our conclusions about the relationship between growth controls and residential building.

First, we considered the effect of simply having any growth control at all (i.e., the fifteen distinct growth controls were lumped together within a "generic growth control" variable), controlling for the state building rate and other distinguishing characteristics of the three study areas (Model One—see Figure A.3). The generic growth control variable assumes a value of 1 for each place and each year in which *any* growth control is in effect and a value of 0 for all other cases.

Next, we took into account further variation between localities by including all the place variables instead of grouping them by study area. In this second model, we still considered only the generic growth control variable rather than the specific measures. Again, we controlled for the state building rate (Model Two—see Figure A.4). This model tests the significance of having any growth control at all, but controls more specifically for other unique differences between jurisdictions, including variations within each study area.

Finally, we introduced the fifteen distinct growth control measures as possible predictors of building activity in place of the generic growth control variable, holding constant the differences between the study areas and the state building rate (Model Three—see Figure A.5). At each step along the way we tested the statistical significance of including or excluding variables. We found that including all of the place variables rather than just grouping them by study area as in the previous models did not increase the explained variance significantly, though it clarified the sources of variation. Including the distinct growth controls likewise did not significantly increase the amount of variance explained but showed us how growth controls have diverse impacts. In Model Three, we no longer control for variation between places within the study areas, but this model provides a superior test of growth controls, since it measures the impact of each distinct regulatory measure.

The Substantive Significance

The overall pattern is quite clear: Broad business trends, as reflected in statewide building rates, are a far more important determinant of local building rates than

FIGURE A.3 Model One

Local Building Activity = Statewide Building Activity + Study Area
Characteristics + Generic Growth Control +
Unexplained Variation

Study Area Characteristics = Local Characteristics Grouped by Study Area
Generic Growth Control = Cases Where *Any* Growth Control Is in Effect

FIGURE A.4 Model Two

Local Building Activity = Statewide Building Activity + Local Characteristics +
Generic Growth Controls (Any Growth Control) +
Unexplained Variation

FIGURE A.5 Model Three

Local Building Activity = Statewide Building Activity + Study Area Characteristics + Local Growth Controls + Unexplained Variation the enactment of growth controls. This held true for all of the models we considered and for all lag times we introduced. For every percentage point change up or down at the state level, local building rates change almost a full point (.85) in the same direction (p<.01). The estimated effect of state building rates varies slightly between our various regression equations; see Table A.1 for coefficients and t-ratios for each of the models. Again, the data we report here are for a two-year lag.

Only three of the fifteen growth controls had a significant impact on subsequent building activity, and one of these was in the wrong direction. Residential building rose a significant 6 percent (p<.05) in Corona after a 1986 voter initiative increased the minimum residential lot. The two other statistically significant controls were the earliest measures in the Santa Barbara study area: a 3.7 percent decrease in building activity followed the 1972 Goleta water moratorium (p<.01), and a 1.6 percent decrease followed the 1973 Santa Barbara city downzoning (p<.05). These data confirm some earlier research on the Santa Barbara case, which suggested that these strong measures did indeed have a detectable effect on housing construction.[4]

To walk through this step-by-step, our simplistic statistical model (Model One) suggested that having *any* growth measure at all reduced local building activity regardless of the broader development trends within the state. But when we introduce, as is appropriate, more specific local differences (Model Two), the "generic" impact of growth control disappears. What appeared to be a dampening in building activity produced by having any growth control policy was instead a sign of overarching differences between the particular jurisdictions. In Model Three, we tested the effects of specific growth controls on building activity and learned that few have any significant bearing on subsequent building activity.[5] When we measured the impact of growth controls using a three-year lag, all the results were virtually identical with a two-year lag. With a single-year lag, none of the growth controls significantly decrease the amount of development. Only the acceleration of building activity that followed the 1986 Corona regulatory change (mentioned above) remains significant in this case.

When we use a fine-grained analysis that distinguishes the impact of specific measures on local residential building rates and apply appropriate statistical tests, we find that the dampening effects of growth controls have been exaggerated greatly.

TABLE A.1
Predicting Local Residential Building Rates: Coefficients and (t-ratios)

	Model One		Model Two		Model Three	
Places						
Santa Barbara (default)						
Carpintería	-	-	2.01	(1.4)	-	-
Goleta	-	-	0.7	(0.9)	-	-
Santa Monica	-	-	0.14	(0.21)	-	-
City of Riverside	-	-	1.68	(3.23)[b]	-	-
Corona	-	-	4.6	(2.44)[a]	-	-
Hemet	-	-	3.8	(6.24)[b]	-	-
Lake Elsinore	-	-	4.22	(2.44)[a]	-	-
Norco	-	-	6.12	(1.71)	-	-
Perris	-	-	7.44	(2.17)[a]	-	-
Western Riverside County	-	-	3.4	(3.65)[b]	-	-
Study Areas						
Santa Barbara (default)						
Riverside Study Area	2.9	(4.85)[b]	-	-	1.81	(2.75)[b]
Santa Monica Study Area	−0.42	(−.076)	-	-	−1.55	(−2.21)[a]
Growth Controls						
Santa Barbara 1	-	-	-	-	−1.61	(−2.05)[a]
Santa Barbara 2	-	-	-	-	−0.24	(−0.3)
Santa Barbara 3	-	-	-	-	−0.92	(−1.1)
Goleta 1	-	-	-	-	−3.73	(−5.44)[b]
Goleta 2	-	-	-	-	1.33	(1.9)
Goleta 3	-	-	-	-	0.04	(0.05)
Santa Monica 1	-	-	-	-	−0.67	(−0.84)
Santa Monica 2	-	-	-	-	−0.37	(−0.42)
Santa Monica 3	-	-	-	-	0.39	(0.42)
Riverside 1	-	-	-	-	−2.13	(−1.1)
Riverside 2	-	-	-	-	0.84	(0.37)
Riverside 3	-	-	-	-	−0.27	(−0.11)
Corona 1	-	-	-	-	−0.4	(−0.16)
Corona 2	-	-	-	-	0.82	(0.43)
Corona 3	-	-	-	-	6.1	(2.18)[a]
Generic Growth Control	−1.27	(−2.96)[b]	−0.6	(−1.49)	-	-
State Building Rate	0.89	(5.0)[b]	0.93	(5.51)[b]	0.85	(4.75)[b]
Log Likelihood Ratio	−497.53		−506.22		−502.86	

NOTE: Model 1 Includes study areas and generic growth control; Model 2 includes all places and generic growth control; Model 3 includes study areas and specific growth control; All models calculated with two-year lags.

[a] Significant at .05.

[b] Significant at .01.

Notes to
Appendix A

1. We were forced to exclude Moreno Valley because it was incorporated in 1986; construction data thus does not exist for the prior period. Likewise Temecula, not incorporated until 1989, is excluded.

2. Initially, we used a lag time of one year, following prior research and in accord with the California Permit Streamlining Act of 1980, which requires localities to process development applications within one year or to automatically grant approval. In response to reviewers' comments, we performed additional analyses for two- and three-year lag periods. General results of all three time lags will be discussed below.

3. An exception was made for one restriction, the 1977 Corona building permit moratorium, which we presumed would have an immediate effect.

4. Prior investigations that attribute significance to the Goleta water moratorium are Appelbaum, 1978; Mercer and Morgan, 1982; and Rey, 1988. Rey (1988) found the Santa Barbara downzoning significant, but Appelbaum (1986) found it insignificant.

5. Even here, it may be that specific growth controls appear significant only because of local differences within the study areas. After all, the "generic" significance of growth control washed out when we moved from Model One to Model Two and replaced the study area variables with the specific place variables.

Appendix B:
Chronologies of Growth Control

Santa Barbara Chronology

The City of Santa Barbara

1967   May: *Harbor Expansion*—City voters turn down $1.5 million bond issue to expand the city harbor. Vote 11,310 to 5,036.

1971   *Citizens General Plan Goals Committee Report*—Goals are adopted as part of the city's general plan recognizing that all major elements of the community's economic base (pension and property income, tourism, R&D) depend directly on maintaining and enhancing the quality and character of Santa Barbara. The goals call for a study of the impacts of growth.

1974   *Impacts of Growth Study*—A planning task force provides research for community analysis and discussion of optimum population size. The studies identify the ideal city size as 50,000–100,000, emphasizing that there must be a balance between residential and nonresidential growth.

1974   November: *Wilcox Property*—City residents vote 13,422 to 11,041 to support city purchase of seventy-acre Wilcox ocean bluff-top property for open space or park to preclude the construction of eighty-three homes on the site. Vote was advisory, and the city declines to purchase the property.

1975   *Residential Downzoning*—The city amends the general plan and zoning to reduce allowable densities in residential areas and to set a population goal of 85,000.

1977   March: *85,000 City Population Ceiling*—Voters endorse 85,000 limit stated in the general plan; this passes with a vote of 11,734 to 7,110. A second measure, requiring that changes to the general plan and zoning ordinance be approved by the voters, passes with a vote of 10,950 to 7,330.

1979   March: *City Park Lands*—City residents vote against disposing of certain city park lands.

1982   *Bullock's Department Store*—City residents defeat a measure that would have permitted a Bullock's department store to replace several existing local establishments on State Street.

1982   November: *Living Within Our Resources*—City residents vote in favor of a charter amendment (Measure K) that requires that the city's "land development shall not exceed its public services and physical and natural resources." The measure stipulates that future rezonings require a "supermajority" (five of seven votes) for city council approval.

1982   *Revisions to Development Plan Review*—The city institutes a new approval process for nonresidential projects larger than 10,000. They are to be approved at the discretion of the planning commission according to the projects' conformance with "sound community planning," compatibility with "neighborhood aesthetics/character," and impact on the availability of housing.

1985   March: *Red Lion Inn*—City voters approve a permit to construct Fess Parker's Hotel on the city's waterfront.

1987   November: *Wilcox Property*—More than 60 percent of Santa Barbara voters favor public purchase of the property, but the measure falls short of the two-thirds majority required to pass the bond issue. The vote is 63.4 percent to 36.6 percent.

1983–1987 *Phases I and II of the General Plan Update*—Technical studies are done to evaluate the appropriate balance between residential and nonresidential growth. Workshops are held to involve the community in designing longer term limits on growth—specifically nonresidential growth.

1986   *Interim Water Moratorium*—An interim ordinance is passed barring the city from approving new development projects that require new water. The city is charged with seeking out new water sources.

1988   June: *Wilcox Property*—A second attempt to approve a bond issue to buy the property fails, falling short of the required two-thirds majority of city voters. The vote is 60.5 percent in favor and 39.5 percent opposed.

1988   December: *Long-Term Water Ordinance*—The city continues the suspension of development permits requiring new water. At the same time, the city establishes a program to allocate fifty acre feet per year for residential development, human services, government, and hardship cases.

1989   *Housing Mitigation Program*—A housing mitigation program is enacted, requiring developers to replace any housing units that they demolish and to provide for the "new demand units" created by a project through direct construction or in-lieu fees.

1989   November: *Commercial Growth Limit*—Ballot Measure E passes, limiting commercial development to 3 million square feet over the next twenty years. The measure passes with a vote of 11,784 to 9,285.

1990   February: *Stage III Drought*—The city declares a Stage III drought, putting severe restrictions on water use and introducing a tiered pricing system.

Santa Barbara County

1970   November: *Voters deny El Capitan Ranch Subdivision* —Voters rescind approval of the subdivision of 3,638 acres next to El Capitan State

Beach with a vote of 46,861 to 33,822. The plan, which would have allowed 1,535 homes, had the unanimous support and approval of the county planning commission and the board of supervisors.

1972   *Goleta Water Moratorium*—Board passes a water moratorium denying new hookups within their district, which covers the western suburban areas of the South Coast.

1973   May: *Voters Support the Goleta Water Moratorium*—Two measures concerning the water moratorium are considered by Goleta voters. The measure to continue the moratorium passes by 11,107 to 5,006, while an opposing measure to rescind the moratorium is defeated 12,220 to 3,791.

1974   February: *County Establishes Office of Environmental Quality (OEQ)*— The task of this office is to carry out initial environmental assessments and prepare EIRs. The county abandons the practice of accepting EIRs prepared by developers. The OEQ is given the final word on environmental review.

1975–1976 *Pro-Growth Grand Jury Slams the OEQ and Its Director*—Growth advocates suggest that OEQ be merged with planning and that the board of supervisors oversee the EIR process.

1977   *Supervisor Initiates Consideration of Growth Management Ordinance for Goleta*—Supervisor Yager forms a citizen committee to design a growth management ordinance. The ordinance aims at limiting residential permits to a yearly rate of 1.2 percent.

1978   June: *County Voters Approve Population Growth Rates for South Coast*— County voters pass two advisory measures (O and P) proposing a population growth rate of .9 percent per year. Planning staff recommends a point allocation system but puts off decisions, pending study of the impacts of commercial growth on population.

1977   November: *County Creates Department of Environmental Resources*— The supervisors elevate the OEQ to full department status, maintaining the autonomy of environmental review, but also making the department head a political appointment, rather than civil service. The past director is kept on but later fired by the board of supervisors.

1979   March: *County Voters Reject State Water*—Voters turn down a bond proposal to finance a pipeline for importing water from Northern California by a vote of 43,846 to 17,301.

1980   *County Completes Study of Commercial Growth Impacts*—The county completes a Regional Growth Impact Study(RGIS). Staff recommends a permit allocation system to control growth and achieve a jobs/housing balance. Supervisors propose that housing permits be granted at an annual rate of 1.8 percent to achieve a .9 percent population growth rate. They target a maximum of 279,500 square feet of nonresidential approvals for the first year, including commercial, industrial, and government (CIG).

1981   *County Supervisors Decide to Incorporate Growth Management into the Comprehensive Plan*—The supervisors delay action on a growth management ordinance with the idea that it should be incorporated into a revised comprehensive plan.

1981   December: *County Staff Recommends Changes to Comprehensive Plan*—
        The staff proposes policy changes for adding growth management to
        the comprehensive plan. The supervisors approve three of the five
        policies but "leave the teeth out" by not setting a cap on the number
        of residential approvals.

1983   *Growth Impact Study Updated*—The county updates research on the
        impacts of commercial growth (IGIS–1).

1984   November: *Recall of Slow-Growth Water Board Fails*—An attempt to re-
        call the slow-growth majority of the Goleta Water Board fails.

1985   *Hyatt Hotel Project Approved*—County supervisors approve Hyatt pro-
        posal in a 4–1 vote. The project, which would ultimately include 524
        rooms, is planned for an undeveloped beachfront loaded with envi-
        ronmental and archeological treasures. The Hyatt is also approved
        unanimously by the California Coastal Commission.

1985   *Second Update of Growth Impact Study*—(IGIS–2).

1985   November: *Pro-Growth Candidates Elected to Water Board*—Two pro-
        growth candidates are elected to the Goleta Water Board, defeating
        slow-growth incumbents. The new majority vows to relinquish re-
        sponsibility for growth management and focus on developing new
        water sources.

1985   November: *Supervisors Limit Population Growth*—The supervisors re-
        spond to the Goleta Water Board by postponing any approvals in Go-
        leta for ninety days and unanimously supporting a target population
        growth rate of .9 percent. Despite their stated intention to limit hous-
        ing approvals to 280 units, the board approves twice that number in
        the following eleven months.

1986   *Board Approves Residential Restrictions for Goleta*—The supervisors ap-
        prove an overlay zone for Goleta by resolution. The Restricted Re-
        sources Overlay District includes a point system for prioritizing pro-
        ject approval. Overall approvals are targeted at 1.2 percent per year to
        achieve desired population growth rate.

1987   November: *Slow Growth Candidates Capture Water Board Majority*—Go-
        leta voters elect a trio of candidates on a slow-growth ticket, shifting
        the majority.

1987   *Board Decides to Include Nonresidential Development in the Growth
        Management EIR*—Public hearings are held in the early months of
        1989.

1987   December: *Court of Appeals Rules Hyatt EIR Invalid*—The court rules
        that Hyatt needs to consider alternate sites and sizing of the hotel to
        satisfy CEQA requirements. Hyatt later appealed to the state supreme
        court and lost.

1988   February: *Hyatt Project Approved by the Planning Commission*—The
        planning commission votes 4–0 to approve a reworked project in-
        cluding 400 rooms.

1988   July: *Economic Analysis of Hyatt Finds That 400 Rooms Are Necessary to
        Project Feasibility*—An independent economic analyst hired by the
        county finds that 400 rooms are needed to make the project viable.

1989   November: *Slow-Growth Candidates Take Over the Board of Supervisors*—Incumbents are defeated in two districts by slow-growth rivals. The entire board is composed of slow-growth representatives.

1989   September: *Ventura Court Rules the Hyatt Has Not Considered Adequate Alternative Sites*—This project is again appealed by the Hyatt with the assistance of the Pacific Legal Foundation—a nonprofit group that defends private property rights (James Watt).

1989   November: *Board Passes Goleta Growth Management Ordinance*—Ordinance limits housing approvals to 200 per year and nonresidential approvals to 80,000 square feet per year.

1989   *Supervisors Initiate Comprehensive Plan Review for Montecito*—The county begins a review of the area plan for Montecito that will include a growth management element.

Santa Monica Chronology

1945   Rental market builds because of Douglas Aircraft employment and war work.

1960   Santa Monica is 69 percent renter.

1966   Santa Monica Freeway is completed.

1970   Demolition of the Santa Monica Pier is opposed by citizen groups.

1973–1975 *Liberal Republicans Elected to Council and New Community Organizations Are Formed*—Christine Reed and Peter Vanden Steenhoven are elected to the city council, and the Ocean Park Community Organization (OPCO) and the Santa Monica Democratic Club (SMDC) are formed.

1975   *Bond for Shopping Center Is Passed*—Citizens sue the city over this project, claiming that it does not return benefits equitably to the community.

1978   January: *Citizen Lawsuit Against Santa Monica Place Settled*—Citizens opposing the development of a publicly assisted commercial mall settle their lawsuit in return for a variety of community benefits to be funded with parking revenues and assisted by a $100,000 contribution by the developer, the Rouse Company. Community benefits include: 20 percent of excess parking revenues, two pocket parks, ten recycling centers, $600,000 of current CDBG funds for housing rehabilitation, $50,000 for child care over five years, $50,000 for women's shelter over five years, and a commitment to obtain CETA staffing for the shelter.

1978   June: *Rent-Control Initiative Fails*–A rent-control measure is spearheaded by Sid Rose, former labor organizer, and sponsored by the Santa Monica Fair Housing Alliance(SMFHA) on behalf of a senior citizen constituency (Campaign for Economic Democracy—CED—was not involved). The measure was defeated 54–46 percent. After the elections Santa Monicans for Renter's Rights(SMRR) is formed by Santa Monica Democratic Club, Campaign for Economic Democracy, and SMFHA.

1979   April: *SMRR Rent-Control Initiative Wins at Polls*—The initiative passes with the city's highest voter turnout yet.

1979   June: *SMRR Wins All Seats on Rent-Control Board*—And two SMRR members (Ruth Goldway and Bill Jennings) are elected to the city council.

1979   November: *Apartment Owners Lose a Counter Rent-Control Initiative*—Another SMRR candidate (Cheryl Rhoden) is elected to the city council.

1980   *SMRR Consolidates*—SMRR is joined by the Ocean Park Electoral Network; Bill Jennings resigns from SMRR and denounces the organization.

1981   April: *SMRR Slate Sweeps City Elections*—SMRR candidates win a 5–2 majority on the city council; the candidates include: Denny Zane (CED), Jim Conn (minister of the church at Ocean Park), Dolores Press (union organizer), and Ken Edwards (Santa Monica Democratic Club). The new council immediately passes Ordinance 1207, calling a six month moratorium on new commercial and residential building in the city (exempting single-family and multi-family projects with 25 percent affordable units, as well as some small-scale commercial in the C–3 zone).

1981   Summer: *Commercial and Industrial Task Force Begins Negotiating*—The task force negotiates developer agreements on major projects that are already "in the pipeline." The job is finished by the city council, with Zane playing a prominent role. The city exacts a variety of social benefits from the Welton Beckett Project (which later became Colorado Place), including funding for parks, day care, and on-site affordable housing.

1981   September: *Council Partially Lifts the Development Moratorium*—The city council passes Ordinance 1220, partially lifting the development moratorium and requiring that all larger projects be reviewed by the planning commission. The planning commission is authorized to be more strict than the current zoning.

1981   October: *New Zoning "Guidelines" Are Established*—The city council passes Resolution 6385, revising what should be allowed in a zone district and essentially reducing the amount of office and commercial space that can be built in the city by 54.5 percent. Mitigation fees, scaled by project size, are established for commercial projects to support arts and social service, affordable housing, traffic and emission abatement, day care and open space.

1981   November: *John Alschuler Hired as City Manager*—Alschuler, who had worked under progressive Mayor Carbone of Hartford, pushes the city to take a proactive attitude toward development, trading new development, particularly hotel development, for such community benefits as jobs and increased tax revenues.

1981   December: *Santa Monica Is Enjoined from Collecting the Mitigation Fees*—Because of suit by the Brotherhood of Carpenters, the city is forbidden from collecting the mitigation fees established in interim procedures. This causes the city to back off on planned zoning changes so that the city's comprehensive plan can first be revised—

	thus establishing firm legal grounding for the zone changes. This suit is eventually settled out of court.
1982	June: *Interim Development Review Procedures Extended*—To allow time for the city to revise the general plan, the city council passes Ordinance 1251, extending the review procedures that had been adopted in September.
1983	January: *Housing Element Adopted*—The city adopts a housing element requiring that housing developments of five and more units include 25 percent affordable units (inclusionary zoning)—this is later reduced to 15 percent. The option of paying "in lieu" fees rather than building affordable units was added in the revised zoning ordinance of 1988.
1983	Spring: *SMRR Loses Majority on City* Council—SMRR loses one seat and fails to unseat two incumbents on the city council.
1984	October: *Land Use and Circulation Element(LUCE) Adopted*—Revision for the comprehensive plan is approved by the city council. The plan was developed with the involvement of community residents and the business community. LUCE reduces the allowable commercial/office square footage by 50 percent. Interim development review procedures are extended again until a new zoning ordinance can be passed.
1984	December: *Council Passes Ordinance 1321*—The city council passes a new zoning ordinance, replacing the interim review procedures.
1986	December: *Final EIR Completed for Revised Zoning*—The scenarios for future development are less restrictive than the earlier proposal for commercial office development, but industrial building potential is radically reduced (by 94–96 percent).
1987	September: *Residential Permit Moratorium Enacted*—The city council passes Ordinance 1416, a moratorium on residential projects in Ocean Park, pending revision of the neighborhood land-use plan.
1988	March/May: *City Council Revises the Fourth Draft of the Revised Zoning Ordinance*—In March, the city council passes interim zoning, pending approval of the new zoning ordinance, to prevent a rush of last minute vesting attempts. In May, the council passes Ordinance 1441.
1988	June: *Final Supplemental EIR Competed for Zoning Ordinance*—In EIR commercial and office space is again reduced by about one third (31.2–37.8 percent) from the last EIR, though industrial space is increased at the same time (47.5–71.3 percent). The net effect on nonresidential square footage is a decrease of 25–32 percent.
1988	August: *New Comprehensive Zoning Ordinance Is Adopted.*
1988	Fall: *Growth Management Proposed for the Ballot*—A growth management ballot issue is proposed but is dismissed on a technicality.
1988	November: *SMRR Candidates Regain Council Majority*—SMRR candidates are elected to the city council; one of the new council members is Ken Genzer, one of the authors of the flawed growth management initiative.
1989	May: *Commercial Moratorium Adopted*—The city council adopts Ordinance 1481, a moratorium on commercial development, because the

city is rapidly approaching levels of commercial development that had been projected for the year 2000 in the LUCE. This ordinance is adopted as an emergency measure for ten and one-half months.

1989   June: *Residential Permit Moratorium Enacted*—The city council passes Ordinance 1484, a moratorium on residential building for the area from Willshire to Montana, pending passage of new development standards.

1989   August: *Initiative to Ban Further Hotel Development on the Beach Qualifies for the Fall Elections*—The sponsoring organization, Save Our Beach, fails to force a special election, falling 300 signatures short of the 8,420 required, but the initiative is slated for the next general elections in November 1990. An alternate initiative is sponsored by the owners of two.hotel projects on the beach, the Sand and Sea Project and the MacGuire Thomas project.

1989   August: *City Council Approves EIR for MacGuire Thomas Hotel Development.*

1990   January: *Airport Redevelopment Thwarted*—The city's plans to redevelop airport land in a public/private venture are stopped by intense neighborhood opposition. The city reverses its October 23 approval of the agreement with a private developer, the Reliance Corporation, to complete a 875,000-square-foot office complex on city-owned land. The group opposing the airport, Santa Monicans for the Public Trust, gathered 9,000 signatures to qualify a referendum putting the project to a public vote.

1990   November: *City-Sponsored Hotel Development Defeated at Polls*—The city's plan to allow redevelopment of the Sand and Sea Club is defeated in city elections.

Riverside Study Area Chronology

1970   *Location of Interstate 15*—Citizens in the city of Riverside oppose the construction of an interstate highway through the city, citing environmental and preservation concerns. The highway is located to the west of the city limits.

1973   *Growth Management System Discussed*—The Corona city council considers a growth management system. The two supporters (Jameson and Rust) are subsequently defeated in their re-election bids.

1976   *Building Permit Moratorium in Corona*—Corona enacts an emergency measure to only issue 450 residential permits for the year, due to limited capacity of sewer and flood control infrastructure. The moratorium is extended through 1977.

1977   *Downzoning of Citrus Land*—The city of Corona downzones 5,000 acres of citrus land south of Ontario Avenue. This formalized what had been an informal policy of the city council, even though the general plan of 1964 slated this area for development.

1977   *Revision of Development Review Process*—Corona adopts a development review process that uses a point system to encourage development adjacent to existing infrastructure.

1977   November: *Riverside City Growth Management Initiative Fails Narrowly*—A ballot initiative (Measure B) to require the city to adhere to its general plan fails by 722 votes.

1979   *Protection of Citrus Groves and Hillsides*—Voters in the city of Riverside pass a ballot measure (Proposition R) to reduce the building densities allowed within agricultural areas of the city and on the hillsides. This proposition responded to city council plans to allow residential development of two-thirds of the citrus area. The measure passes with 66 percent of the vote. Four candidates supporting Proposition R are elected to city council.

1984   *Incorporation of Moreno Valley*—The voters of Moreno Valley elect to incorporate as a city. One of the main motivations is dissatisfaction with county land-use decisions for the area.

1984   October: *Riverside City Council Trades Greenbelt for Park Land Contribution*—The city council approves a residential project at eleven times the density allowed in the greenbelt in exchange for donation of land for Sycamore Canyon Park.

1986   November: *Corona Voters Increase the Minimum Lot Size*—Voters pass Measure H, which increases the minimum lot size to 7,200 square feet for single-family residences. This does not apply to developments that are done through specific plans, which includes most of the major subdivisions (Sierra de Oro, South Corona, etc.).

1986   *Moreno Valley Voters Recall Pro-Development Council Members*—Three of the original members of the city council are recalled because they are doing too much for developers (Horspool makes use of a legal loophole to get re-elected from a district at the same time as he is recalled from his at-large council seat).

1986   *Moreno Valley Building Moratorium*—An eighteen-month building moratorium is imposed on the eastern portion of Moreno Valley. In part this moratorium was imposed to block the private development of an airport and adjacent commercial space by Bengivi-Cohen. The developer withdraws this project when faced with the opposition of all surrounding airports and the FAA.

1987   March: *The City of Riverside Proposes Relaxing Greenbelt and Citrus Restrictions and Grants Exemption to Proposition R*—In the face of landowner protest over Proposition R, the city appoints a blue-ribbon committee and hires a consultant, Peter Dangermond, to study the citrus areas. The consultant suggests that half the area be opened to development—but in a way that would preserve the character of the citrus belt. The city exempts land within the Sycamore Canyon Business Park from Proposition R in return for the donation of fifty acres to the Sycamore Canyon Park.

1987   *The Riverside Voters Strengthen Protection of the Citrus Belt*—Riversiders pass a measure to strengthen Proposition R. Measure C requires a two-thirds majority on the council in order to make exceptions to Proposition R. Controls are also extended to annexed land, and the city is required to evaluate annexations in terms of whether they pay their own way. Measure C passes with 52 percent of the vote, an alternative proposed by the city council (Measure G) is rejected by 62 percent of the voters.

1987   May: *Moreno Valley Establishes Design Guidelines and a Review Board.*

1988   November: *County Growth Management Measure Fails (Measure A)—* This measure would have limited population growth in the county to that of the state, required that local government be able to provide adequate services before approving development, and that open space and agricultural lands be protected.

1989   March: *Riverside City Voters Defeat Measure to Repeal Protection of Citrus Groves and Hillsides—*A proposition (Measure E) to repeal growth controls is defeated by a landslide-76 percent of the voters oppose the measure, despite the over $500,000 spent by the measure's supporters (a city record).

1989   September: *Riverside City Growth Controls Ruled Invalid—*Judge rules Proposition R and Measure C invalid on the grounds that the city's general plan is inadequate. The ruling does not take immediate effect and is appealed by the city of Riverside.

1989   March–December: *County Designs Growth Management Element—*In response to the widespread support for the failed growth control initiative for the county, the county undertakes a planning process to develop growth management policies.

1989   September: *Southwest Area Plan—*The county adopts a community plan for the southwest area. The plan downzones some limited areas on the Santa Rosa plateau but still allows enough development potential for an eventual population of 404,246. They project a population of 120,810 for the year 2010.

1989   November: *City of Temecula Votes to Incorporate—*Voters in the Temecula valley elect to incorporate as a city named Temecula, partly in response to lax county development standards.

1990   June: *Moreno Valley Growth Management Measure Fails (Proposition J)—*A city growth control initiative designed after the earlier county measure is defeated at the ballot box, though it gains 42 percent of the vote.

APPENDIX C Commercial Valuation Data, 1970–1990

| | Total Nonresidential Valuation (in thousands of dollars) | | | | | | | | | | |
	1970	1971	1972	1973	1974	1975	1976	1977	1978	1979	1980
Carpintería	$425	$1,752	$709	$462	$2,637	$474	$1,571	$1,927	$3,162	$3,584	$866
Santa Barbara	$8,488	$2,541	$6,738	$3,273	$5,218	$4,682	$7,655	$12,458	$10,270	$13,927	$10,756
Goleta	$4,787	$2,178	$4,529	$7,664	$5,705	$3,288	$4,589	$8,867	$16,843	$12,230	$9,908
South Coast Balance	$5,361	$6,453	$8,149	$8,571	$7,125	$4,997	$11,442	$11,092	$17,416	$13,624	$11,619
Santa Barbara Study Area	$14,274	$10,746	$15,596	$12,306	$14,980	$10,153	$20,668	$25,477	$30,848	$31,135	$23,241
Santa Monica	$11,117	$10,970	$11,841	$9,823	$8,101	$15,244	$10,646	$17,915	$30,063	$105,804	$59,638
City of Riverside	$19,094	$12,185	$22,908	$30,910	$16,710	$12,849	$16,363	$30,324	$35,041	$48,987	$75,278
Corona	$4,434	$3,262	$4,460	$5,479	$1,552	$4,141	$5,077	$11,218	$13,133	$19,232	$11,135
Hemet	$1,585	$5,077	$4,808	$4,411	$3,708	$3,096	$4,757	$15,464	$5,477	$8,380	$16,406
Lake Elsinore	$107	$736	$660	$1,255	$144	$211	$1,468	$595	$845	$1,159	$4,194
Moreno Valley											
Norco	$734	$774	$740	$1,068	$671	$611	$1,847	$2,645	$2,641	$4,569	$3,107
Perris	$845	$1,532	$316	$783	$493	$1,361	$771	$1,463	$561	$630	$1,573
Western Unincorp. County	$10,097	$19,446	$24,826	$19,413	$11,957	$11,416	$22,676	$23,033	$27,460	$71,666	$69,438
Study Area	$17,802	$30,827	$35,810	$32,409	$18,525	$20,836	$36,621	$54,418	$50,117	$105,636	$105,853
California	$2,543,764	$2,969,469	$3,048,634	$3,230,637	$3,258,723	$3,106,221	$3,467,649	$4,712,638	$5,958,365	$7,294,901	$7,470,107
U.S.	$13,224,000	$15,005,000	$17,289,000	$20,903,000	$20,483,000	$16,200,000	$19,182,000	$23,663,000	$32,200,000	$38,300,000	$34,800,000

(continues)

167

APPENDIX C Commercial Valuation Data, 1970–1990 (continued)

	Total Nonresidential Valuation (in thousands of dollars)									
	1981	1982	1983	1984	1985	1986	1987	1988	1989	1990
Carpintería	$4,072	$706	$1,239	$5,920	$3,532	$2,403	$3,851	$5,639	$4,113	$3,815
Santa Barbara	$10,967	$13,274	$15,309	$40,377	$25,958	$26,527	$23,046	$21,194	$42,076	$16,676
Goleta	$21,614	$14,532	$20,558	$32,482	$13,081	$44,646	$28,104	$29,663	$21,111	$18,968
South Coast Balance	$22,251	$16,200	$23,848	$33,144	$27,130	$47,564	$32,774	$32,760	$21,719	$42,514
Santa Barbara Study Area	$37,290	$30,180	$40,396	$79,441	$56,620	$76,494	$59,671	$59,593	$67,908	$63,005
Santa Monica	$80,787	$65,238	$21,181	$49,068	$57,258	$95,358	$48,284	$127,752	$142,869	$123,070
City of Riverside	$62,794	$51,388	$43,730	$88,664	$88,554	$107,830	$107,819	$92,172	$115,763	$122,482
Corona	$9,742	$8,954	$23,913	$39,170	$32,752	$38,394	$69,036	$76,378	$102,738	$71,609
Hemet	$17,358	$15,917	$10,774	$10,615	$13,028	$13,890	$7,151	$7,969	$23,360	$13,454
Lake Elsinore	$1,210	$1,506	$5,053	$3,022	$6,358	$6,979	$12,816	$10,844	$14,645	$7,919
Moreno Valley					$109	$19,066	$35,413	$48,078	$48,476	$58,650
Norco	$1,324	$2,736	$5,209	$2,737	$2,189	$3,023	$2,529	$1,550	$2,739	$150
Perris	$584	$1,154	$1,477	$5,096	$5,678	$6,116	$6,415	$11,289	$10,206	$9,978
Western Unincorp. County	$80,968	$93,891	$91,680	$103,441	$102,782	$148,730	$98,659	$143,726	$184,528	$188,353
Study Area	$111,186	$124,158	$138,106	$164,061	$162,896	$236,198	$232,019	$299,894	$386,692	$350,113
California	$8,905,843	$8,729,875	$9,910,801	$11,973,109	$13,317,778	$13,194,906	$12,259,547	$13,375,697	$13,666,100	$12,735,540
U.S.	$39,300,000	$34,800,000	$37,000,000	$47,200,000	$52,700,000	$50,600,000	$50,000,000	$51,544,000	$52,271,000	$47,529,000

SOURCES: Construction Industry Research Board, 1971–1990; Riverside County, 1970–1989; Santa Barbara County, 1970–1989; and U.S. Department of Commerce, 1970–1989.

Appendix D Residential Building Rates

							New Permits/Existing Housing (%)													
	1971	1972	1973	1974	1975	1976	1977	1978	1979	1980	1981	1982	1983	1984	1985	1986	1987	1988	1989	1990
Carpintería	7.81	17.73	10.49	2.75	1.40	1.49	3.16	5.40	1.12	0.68	0.13	0.98	0.20	2.20	1.98	5.38	1.26	0.91	0.42	1.51
Santa Barbara	2.00	2.46	2.39	0.73	0.75	1.14	1.21	0.98	0.74	0.90	0.54	0.50	0.45	1.08	0.89	1.12	1.61	0.31	0.15	0.16
Goleta	6.49	7.46	7.39	0.47	0.09	0.13	0.25	0.18	0.24	0.18	0.18	0.73	1.38	1.24	1.81	0.93	0.99	0.73	0.98	0.57
South Coast Balance	7.82	9.11	9.43	1.67	0.56	0.80	0.97	0.70	0.48	0.48	2.25	0.76	1.37	1.23	1.80	1.43	1.45	1.13	1.46	0.93
Santa Barbara Study Area	4.18	5.49	5.31	1.20	0.72	1.04	1.25	1.17	0.67	0.73	1.12	0.63	0.76	1.21	1.29	1.52	1.53	0.65	0.65	0.54
Santa Monica	3.64	4.34	4.87	1.11	0.56	0.78	0.67	1.45	1.15	0.94	0.39	0.44	0.10	0.04	0.49	0.29	0.65	0.87	0.72	0.58
City of Riverside	4.11	3.55	3.29	1.44	1.58	4.42	10.65	0.28	1.07	0.33	0.97	0.29	1.85	4.93	4.97	4.42	2.08	2.12	2.92	1.59
Corona	3.21	8.54	3.90	2.52	5.64	6.90	1.54	1.32	2.44	3.15	3.01	0.75	3.31	6.86	6.34	12.46	10.92	19.01	13.31	4.79
Hemet	5.62	7.66	4.61	2.90	5.19	5.16	6.37	7.04	7.92	1.46	1.09	1.79	5.89	4.90	5.67	13.01	2.69	4.19	3.80	8.79
Lake Elsinore	0.55	1.97	1.84	1.66	1.82	2.84	2.81	12.45	8.68	12.53	8.17	4.08	8.74	14.43	7.88	7.28	9.12	6.83	4.09	4.39
Moreno Valley																				
Norco	16.94	23.27	15.17	10.02	20.97	23.00	4.72	4.22	3.22	1.07	0.66	0.59	0.79	0.49	2.10	3.49	0.67	0.88	2.92	3.78
Perris	4.03	0.92	3.21	3.52	0.26	0.85	5.95	1.71	16.38	8.40	1.43	5.02	7.54	16.77	5.40	11.77	13.07	38.51	22.02	12.71
Western Unincorp. County	3.88	3.80	3.14	1.93	2.23	4.38	6.90	5.72	5.39	4.14	2.93	2.64	6.93	8.31	7.01	5.24	5.19	13.00	5.98	1.52
Riverside Study Area	4.06	4.36	3.45	1.95	2.61	4.86	7.75	3.40	3.83	2.66	2.17	1.65	4.79	6.97	6.13	7.95	5.51	11.00	7.02	2.79
California	3.67	3.87	2.88	1.66	1.66	2.75	3.29	2.88	2.42	1.62	1.15	.093	1.86	2.38	2.80	3.16	2.45	2.40	2.20	1.47
U.S.	2.79	3.11	2.46	1.42	1.21	1.67	2.13	2.22	1.87	1.41	1.15	1.15	1.83	1.88	1.91	1.90	1.63	1.52	1.37	1.12

SOURCES: Riverside County, 1970–1989; Santa Barbara County, 1970–1989; U.S. Department of Commerce, 1970–1989.

Appendix E:
Case Study Details

In this appendix we present the standards, mitigations, conditions, and fees applied to these projects upon formal approval. We recognize that formal requirements are not always fully enforced (see Chapter 4), but these snapshots provide a basis for comparing local practices.

Santa Barbara Routine Case—Barbara Plaza

Project Context

Date.   September 1990.

Site Description.   This was a 14,066-square-foot site in downtown Santa Barbara, previously used as a used-car sales lot (with a 2,220-square-foot office). It was one block off of the main street of downtown and across the street from the Paseo Nuevo, a downtown mall redevelopment project.

Background.   This was one of the first new commercial projects to come through since the city enacted commercial growth limits by ballot initiative. The planning department selected this project as an example of what routine commercial projects would be like after Measure E. The immediate effect of the measure was to stop new commercial development proposals, except those that were already in process or were somehow exempted from the limits. This measure was enforced immediately by city staff under the direction of the city attorney, because it was a charter amendment, even though it had not yet been translated into zoning law. The size of this project (and its related impacts) were therefore constrained *before* project review, because only certain types of projects could even be considered. This project was proceeding under an exemption that allowed for building additions under 3,000 square feet. The owners could demolish the existing building and retain the right to rebuild this square footage. Combining the original footage with the allowed 3,000-square-foot addition, they ended up with a new building of 5,600 square feet.

In addition to this square footage limitation, project sponsors could use no new water, nor create significant traffic impacts. The developers reduced the size of the project slightly to keep traffic impacts below the threshold of impact. For water, the project tapped the historic water uses on the site and the historic water uses transferred from another site displaced by government action, the Paseo Nuevo redevelopment.

Review Process. The project was smaller than what would have been typical before Measure E. Because it was less than 10,000 square feet, it did not trigger automatic review by the planning commission, though it would have if an environmental impact report had been required. It was still reviewed by the Environmental Review Committee and the Landmarks Committee (design review).

Project Description

A two-story commercial building

Building footprint:	5,975 square feet
Paving coverage:	6,313 square feet
Landscaped area:	2,779 square feet
First-floor retail:	4,700 square feet
Second-floor office:	900 square feet

Standards

Design. The city imposed restrictions on future uses (particularly such traffic generators as restaurants and fast food) that went beyond the zoning definition; city officials required that changes in future uses be reviewed by city staff to assess traffic impacts. All plumbing fixtures had to be ultra-low flow. The Landmarks Committee made some specific suggestions regarding materials and project design, to which the developer agreed.

Conditions/Mitigation

Traffic Impacts. The developer was required to develop a transportation management plan and to incorporate vehicle-use disincentives into building leases. Requirements included the provision of an employee lunchroom with minimum equipment; bus passes for employees; the submission of an annual transportation report to the city; the assignment of an alternative transportation coordinator.

Street and Gutter Improvements. The developer was to make improvements to the streets and gutters adjacent to the site, including handicap access, as approved by the city engineer.

Construction Conditions. Construction traffic was limited and parking was to be provided to workers. Dust was to be controlled by watering during grading.

Archaeological Monitoring. The developer had to contract with a qualified archaeologist and a Native American to monitor grading and demolition to assure the protection of cultural resources.

Fees

Downtown Transportation Improvement Fee. A fee based on 6.68 peak daily trips; $2,350 per daily trip = $15,698.

Affordable Housing. Because of the smaller size of this project, it did not have to mitigate impact on new housing demand; it was not deemed to have a significant impact. The typical commercial project prior to Measure E, at about twice the size, would have (since 1989) been required to provide approximately 4 units of affordable housing.

Santa Barbara Best Case—Rancho***Barbara

Project Context

Date.   May 1986 (as amended).

Site Description.   This was a 28-acre site between the freeway and a major arterial, and adjacent to a shopping mall. It was divided into three development sites, an 8.5-acre site that fronted the freeway and buffered the other two sites of 9.6 acres and 6.9 acres. Part of the site development included bisecting the original parcel of land with a street to connect two major arterials.

Background.   Prior to the adoption of the Rancho Arroyo specific plan in November 1984, the area was used for avocado and citrus groves, though zoned for lower density residential (5 units per acre). The specific plan allowed the construction of auto dealerships along the highway plus higher density housing on the interior parcels—on the condition that a portion of the housing be affordable. In 1986, the project owner got these conditions amended to increase the total number of housing units that would be built.

Site 1 Project Description: Automobile Agencies

	Agency 1	**Agency 2**
Building size	16,207 sq. ft	16,446 sq. ft
(for showrooms, offices, and service facilities)		
Landscaping/easement	39.46%	30.13%
Paving	50.28%	57.51%
Parking required	65 spaces	66 spaces
Parking provided	168 spaces	175 spaces
(with outdoor display area)		

Standards

Water use.   A total of 65 acre-feet per year (AFY) was allowed in the specific plan for the entire project. This portion of the project would use 3 AFY.

Traffic.   Average daily trips (ADTs) set at 936 (6.126 usable acres x 152.8 ADT/acre).

Other Standards
  Textured or colored pavement throughout the project
  Aesthetically pleasing exterior lighting
  Plumbing fixtures had to be water-conserving
  Prohibited use of herbicides and fertilizer within creek drainage

Conditions/Mitigation

Landscaping.   Required to maintain a landscaping plan approved by the Architectural Board of Review. The owner had to provide a performance bond to assure completion of landscaping portion of the plans.

Transportation.
Carpool parking
Free employee bus passes
Employee lunchroom, showers, and lockers
Variable work hours (if feasible)
Customer shuttle service

Construction Conference.   Limitations on impacts of construction traffic, parking, and dust.

Fees.   The project was to pay traffic mitigation fees for 936 ADT to two existing traffic improvement districts.

Site 2 Project Description: Affordable Senior Apartment Project

112 units of housing on 3.5 acres, including four three-story buildings and nine two-story buildings, with a 1,000-square-foot recreation building and a swimming pool.

Building Coverage, 19%   Landscaping/Open Space, 55%
Paving, 26%   Parking, 56 spaces required, 78 provided

Standards

Water.   The specific plan allowed 63 AFY for the entire project. This portion of the project was to use 27.2 AFY.

Underground Utilities.   Utilities had to be underground within the project.

Density.   Housing density was 31.9 units per acre. This exceeded the general plan standard for this area of 12 units per acre but was in line with densities approved for other affordable housing projects within the city. The project was eligible to tap the city's density reserve that was set aside for affordable housing.

Conditions/Mitigation

Landscaping.   Required to maintain a landscaping plan approved by the Architectural Board of Review. The owner had to provide a performance bond to assure completion of landscaping portion of the plans.

Affordability   All units were required to be affordable for thirty years, as described in the specific plan for thirty years. Use was restricted to senior citizens (over 62 years of age) and the number of motor vehicles was restricted.

If the project was not used for seniors, the number of units was to be reduced to comply with city parking requirements.

On-Site Recycling Bins.

Letter of Agreement/Performance Bond.   The owner had to submit a letter of agreement ensuring the enforcement of affordability restrictions. Landscaping was to be assured with a performance bond.

Fees.   The project had to pay traffic mitigation fees for traffic improvement districts based on 351 ADT (112 units x 3.3 ADT x .95).

Site 3 Project Description: Multiple-Family Condominium Project

The project included 96 condominiums on 9.65 acres of vacant land. 46 units were affordable and 50 were market rate. There were eighteen one-story and two-story buildings, with clusters of 5–6 units, as well as three free-standing units and one duplex. A small cabana and swimming pool and two tennis courts were provided.

Building Coverage, 21%	Landscaping/Open Space, 62%
Paving, 17%	Parking, 215 spaces (192 covered)

Standards

Density.   The overall density was 9.9 units per acre, less than the general plan designation of 12 units per acre. Setbacks of twenty to thirty feet were provided along the west property line, thirty to forty feet along the south property line, and forty to sixty feet along the east property line.

Other Standards.   Existing trees on the property were to be preserved, protected, and maintained.

Textured or colored pavement
Underground utilities within project site
Plumbing had to be water-conserving

Conditions/Mitigation

Landscaping.   The developer had to maintain a landscaping plan approved by the Architectural Board of Review and had to provide a performance bond to assure completion of the landscaping portion of the plans.

Affordability.   46 three-bedroom units were to be sold at prices affordable to low-moderate income households. Resale prices were controlled through recorded documents to maintain affordability for thirty years.

On-Site Recycling Bins.

Letter of Agreement.   The owner had to submit a letter of agreement ensuring the enforcement of affordability restrictions.

Fees. The project had to pay traffic mitigation fees for traffic improvement districts based on 499 ADT (96 units x 5.2 ADT).

Santa Monica Routine Case—Monica Tower

Project Context

Date.   November 1988.

Site Description.   The project was located on the edge of downtown Santa Monica on the corner of two major arterials. The site was approximately 30,000 square feet.

Project Description. The project was a six-story, 93,451-square-foot retail/bank/office development including 312 parking spaces in a four-level subterranean garage. Approximately 80 percent of the proposed building area was offices and 20 percent was for retail and a bank. The project design conformed with the specifications of the Third Street Specific Plan in terms of height, bulk, use, and urban design policies.

Standards

Design.   The maximum building height was 80 feet, 6 inches, but the building stepped back on the upper floors to reduce the overall mass and bulk. The developer was required to reduce each of the top floors by 1,000 square feet to reduce bulk. Other design features such as awnings and inset windows were intended to break up the scale of the building. There was a landscaped public courtyard on the ground level. The developer was required to use nonreflective glass. The overall floor area ratio was 3.11.

Traffic.   The project provided enough parking on-site that it did not affect the availability of downtown parking. The project alone did not significantly impact the level of service at any of the eight intersections analyzed in the EIR.

Conditions/Mitigation

Transportation.   Owners were required to encourage public transit usage, provide bicycle parking, and construct a bus shelter.

Parking.   The owner was allowed to charge for parking but had to provide free valet parking during all hours of operation.

Wastewater.   Developers had to retrofit existing Santa Monica residences with ultra-low flow plumbing so that the new development would not increase wastewater flows (maximum cost, $30,000). Ultra low-flow plumbing fixtures were required throughout.

Landscaping.   The public courtyard had to include 60 percent plant materials.

Construction.   The developer was required to try to reduce dust by watering, and so on, but there were no specific limits on when construction activities could take place.

Fees

Parks/Housing. The developer had to pay a housing/parks mitigation fee of $2.25 per square foot for the first 15,000 square feet of net rentable office floor area and $5.00 per square foot for the remaining office space, adjusted for inflation. The unadjusted fee for this project based on an office floor area of 76,025 square feet was $338,875. The developer could elect to provide housing and parks directly.

Transportation. Transportation fees could be imposed at some later date.

Santa Monica Best Case—Monica***Centre

Project Context

Date. March 1988.

Site Description. The property was in a semi-industrial area of Santa Monica, just north of the Santa Monica Freeway. The total area was 740,928 square feet (17.1 acres) within the Special Office District.

Project Description. This was a mixed-use commercial project that included four major buildings (referred to as North, West, South, and East), each with 314,894 square feet and with a maximum height of six stories (eighty-four feet). The project included underground parking of 3,652 parking spaces (2.9 spaces per 1,000 square feet of floor area). The development was to be built in two phases but was submitted as one package to lock in development rights.

Standards

Floor Area. The total floor area was not to exceed 1,259,577 square feet, 1.7 times the area of the property.

Building Setbacks. The buildings had to be set back from the street as follows: North—30 to 103 feet; West—45 to 103 feet; South—20 to 101 feet; East—35 to 90 feet.

Design. The design was to be approved by the city's Architectural Board, except for those issues such as building height and floor area that were specifically addressed in the development agreement.

Uses. Land uses were allowed according to the Special Office District but with these maximum limits (which could be appealed):

Restaurant, 50,000 square feet
Medical office, 20,000 square feet
Retail serving the project itself, 40,000 square feet
Health club, 25,000 square feet
Banks and savings and loans, 30,000 square feet
General commercial office, the remainder of the
  allowable floor area for the project

Open Space.   The project was to include "ample" open space and landscaping as specified in the landscaping plan, including a large artificial lake.

Water Conservation.   The project included water saving measures, such as watering at night, using low-flow fixtures, and drought resistant plantings. The goal was to reduce water use by one-third.

Conditions/Mitigation

Phasing.   The project had to be built in no more than two phases. The first phase was to include South, East, and the lake. The second phase of the project had to begin within twelve years of the agreement.

On-Site Child Care.   The developer was required to provide a facility for child and infant care including 3,500 square feet of interior space and 3,500 square feet of exterior space (not by a public street). The city retained the right to approve the child care provider and to require that the provider be a nonprofit organization. Ten percent of the child care capacity was to be reserved for people in the immediate neighborhood with demonstrated financial need at 60 percent of the market rate.

Transportation.   The developer had to submit a traffic demand management (TDM) program and designate a full-time employee to promote and manage this program. The goal of the program was to reduce auto trips by 20 percent. The project's yearly traffic mitigation fee was to be reduced to the degree that the TDM goals were achieved. The TDM programs had to be written into leases on the building to promote tenant involvement. The TDM program was expected to include:

Flex-time
Car-pool parking
50 percent discount tokens for public transportation
Vans and employee shuttles

College Parking.   The developer agreed to provide overflow parking for Santa Monica Community College for a fee and to the degree this was feasible.

Employment.   The developer committed to design and promote job training programs to meet the needs of surrounding neighborhoods.

Wastewater.   The development agreement did not override sewer moratoria.

Preferential Parking Study.   If the city deemed it necessary, the developer would be required to finance a study to see whether preferential parking was needed in the immediate vicinity.

Fees

Housing and Parks Fee.   The developer had to pay a mitigation fee based on $2.25 per square foot for the first 15,000 square feet of net rentable floor area and

$5.27 per square foot for everything above, indexed for inflation. The total fee, not adjusting for inflation and assuming that the maximum amount of retail space ended up being built, would have been $5,828,520. The developer was required to pay at least $3.5 million of this fee before commencing Phase 1.

Traffic Demand Management Fee. The developer had to pay $200,000 per year adjusted by inflation. This was reduced to the extent that traffic demand was reduced by more than the goal of 20 percent in a year. The developer was required to pay for a traffic engineer to prepare an annual analysis of actual traffic so that TDM goals could be evaluated.

Traffic Improvement Fee. The developer had to pay a total fee of $6,408,486 for traffic improvement. $3,950,000 had to be paid prior to receiving the building permit for Phase 1, and the balance before Phase 2. These funds were to be used to reduce traffic levels, including up to $500,000 for staff costs.

Art. The developer had to pay $150,000 to the Santa Monica Arts Foundation for artwork to be placed in public places in Santa Monica.

Homeless Services. The developer had to pay the city $300,000 to be used for services to homeless persons or for the operation of shelters.

Riverside Routine Case—Riverside Arms

Project Context
Date. June 1990.

Site Description. A 4.5-acre site, a portion of which was previously developed with single-family residences (proposed for removal). Zoned R–1, single-family, with proposed zone change to R–3, multi-family, with thirty-foot height limit.

Project Description. Apartment complex with 102 units: 46 one-bedroom and 56 two-bedroom units—22.6 units per acre.

Review Process. Recommended by planning staff with conditions; the design review board looked at building and landscaping plans. The planning commission approved the project with a vote of 7–1.

Standards

Project Design. Variances allowed reduction of the required setbacks from side property lines from fifty to forty feet, to make the rear yards smaller than zoning code, to eliminate the requirement for recreational vehicle parking, and to reduce the neighborhood common space. No exposed mechanical systems (e.g., heating and cooling) were allowed on the pitched roofs.

Conditions/Mitigation

Street trees and parkway landscaping specifications to screen parking area
Trash enclosures of masonry block with stucco finish to match the building

Curbs and gutters

Landscape easement for ten-foot planter on street frontage
  that was to be maintained by applicant

Deed to allow the city to widen adjacent street

Approval did not guarantee sewer availability

Off-site improvements to be specified by Public Works Department

Fees. Not specified in this document.

Riverside Best Case—Riverside***Place

Project Context

Date.   October 1990.

Site Description.   This was a 121-acre site bordering a freeway. The site was undeveloped except for one single-family residence. It was surrounded by single-family residential zones on three sides, which were partially developed, and on one corner abutted a residential-agriculture zone. To the northwest it bordered commercial zoning along the railroad right-of-way and the freeway.

Project Background and Description.   The land was owned by the community college, which had planned to build a new campus. Instead, they teamed up with a private developer who planned to build a conference center for the college and a mixture of office, hotel, and apartments. The developer originally proposed 1,050 apartments; the planning department recommended 517. The planning commission approved the project 7–1 on October 4, 1990, reducing the housing units to 756. Eight hundred people attended the commission hearing, mostly opponents; this was the largest turn-out commissioners could remember in recent history.

Standards (from planning department recommendations)
  570 residences

1. 53 units on 7.5 acres at 7 dwelling units per acre

2. 517 units on 27 acres at 15 dwelling units per acre with thirty- to forty-foot height limits

   400,000 square feet of offices on twenty acres with six-story height limit (seventy feet)

   100,000 square feet of offices on 6.5 acres with two-story height limit (thirty feet)

   260,000 square feet of commercial/retail on twenty acres with two-story height limit (thirty feet)

   250,000 square feet of hotel and a 50,000-square-foot conference center on 8.8 acres with two-story height limit (thirty feet)

Buffering from neighboring residential areas to the northwest, including twenty-five feet of landscaping then a six-foot wall; a fifty-foot building setback from the street; and a two-story height limit (thirty feet) within one hundred feet of the street. The rest of the high-density area would be limited to three stories (forty feet)

Eighty-foot building setback for commercial/office area on street frontage with fifteen feet of landscaping

Lower-density housing to abut the residential areas to the south and the greenbelt (floodplain), or eliminate vehicular access and impose twenty-foot height limit

Conditions/Mitigation

Developer had to pay full cost of traffic improvements adjacent to the site, including street widening and signaling. Developer was to pay one-third of the approximately $10 million cost of widening the bridge over the freeway

100-space park and ride within commercial area

Dedication of streets and a bus stop

Creation of a landscape maintenance assessment district instead of deeding landscaped areas to the city

Transportation management plan, included measures such as carpool preferential parking, bikeways, and van pool incentives, to mitigate air pollution

Solar power for water heating

Research on the historic ranch complex was to be presented to the Historic Resources Department

Cultural resources monitor to be hired by the developer, with the power to halt grading for archaeological evaluation

Developer required to pay pro rata share of new fire station

Water conservation measures—a number of conservation techniques were recommended, including ultra-low-flush toilets

Fees

Park fees
Library fees

Appendix F:
Interview Schedule

These are some areas we would like to discuss for each project:

Is it likely that this project would be approved in your city? If not, how would it have to be altered in order to be approved? If likely, is this a typical project?

Which of the conditions imposed would *not* be imposed on this project in your city? Are these conditions imposed on any projects? What projects? How would you explain the difference?

What other conditions would you require of this project? Would you implement conditions differently? Why? How?

How would the project review differ? Would the project go to the city council? Planning commission?

How do the development impact fees compare? Would you assess different kinds of fees? Different amounts?

Reference List

Advisory Commission on Regulatory Barriers to Affordable Housing. 1991. *"Not in My Back Yard": Removing Barriers to Affordable Housing.* Washington, D.C.: U.S. Department of Housing and Urban Development.

Albrecht, Don E., Gordon Bultena, and Eric Hoiberg. 1986. "Constituency of the Antigrowth Movement: A Comparison of Urban Status Groups." *Urban Affairs Quarterly* 21 (4): 607–616.

Alford, Robert, and Roger Friedland. 1985. *Powers of Theory: Capitalism, the State, and Democracy.* Cambridge: Cambridge University Press.

Almeida, Paul. 1994. "The Network for Environmental and Economic Justice in the Southwest: Interview with Richard Moore." *Capitalism, Nature, and Socialism* 5 (1): 21–54.

Alterman, Rachelle, ed. 1988. *Private Supply of Public Services: Evaluation of Real Estate Exactions, Linkage, and Alternative Land Policies.* New York: New York University Press.

Appelbaum, Richard P. 1978. *Size, Growth, and U.S. Cities.* New York: Praeger.

_____. 1986. "Regulation and the Santa Barbara Housing Market." Report to the California Policy Seminar. Sacramento: California Policy Seminar.

Appelbaum, Richard P., Jennifer Bigelow, Henry Kramer, Harvey Molotch, and Paul Relis. 1976. *The Effect of Urban Growth: A Population Impact Analysis.* New York: Praeger.

Appelbaum, Richard P., and Perry Shapiro. 1983. "Analysis of the Downtown Retail Revitalization Project." Santa Barbara, Calif.: Filed with City Redevelopment Agency.

Aronson, Hal. 1993. "Becoming an Environmental Activist: The Process of Transformation from Everyday Life into Making History in the Hazardous Waste Movement." *Journal of Political and Military Sociology* 21 (1): 63–80.

Associated Press. 1991. "Red Tape Hamstrings Home-Buyers." *Santa Barbara News Press,* July 9, A4.

Association of Bay Area Governments. 1980. *Development Fees in the San Francisco Bay Area.* Berkeley: Association of Bay Area Governments.

_____. 1984. *Development Fees in the San Francisco Bay Area: An Update.* Berkeley: Association of Bay Area Governments.

Babcock, Richard F. 1969. *The Zoning Game.* Madison: University of Wisconsin Press.

Bachrach, Peter, and Morton Baratz. 1962. "The Two Faces of Power." *American Political Science Review* 56 (4): 947–952.

Baldassare, Mark, and William Protash. 1982. "Growth Controls, Population Growth, and Community Satisfaction." *American Sociological Review* 47 (June): 339–346.

Banfield, Edward C. 1961. *Political Influence.* New York: Macmillan.

Banham, Reyner. 1971. *Los Angeles: The Architecture of Four Ecologies.* New York: Harper & Row.

Bartik, Timothy J. 1988. "The Effects of Environmental Regulation on Business Location in the United States." *Growth Change* 19 (3): 22–44.

Battle, Virginia, and Jack Underhill. 1986. "Coming to Grips with the U.S. Enterprise Zone Experiment: A Summary of Ten Case Studies." *Enterprise Zone Notes* (U.S. Department of Housing and Urban Development) (fall): 11–14.

Bauman, Gus, and William H. Ethier. 1987. "Development Exactions and Impact Fees: A Survey of American Practices." *Law and Contemporary Problems* 50 (1): 51–68.

Bean, Lowell J., and Charles R. Smith. 1978. "Gabrielino." In *Handbook of North American Indians*, Vol. 8, *California*, ed. Robert F. Heizer, 538–549. Washington, D.C.: Smithsonian Institution.

Beatley, Timothy, and Kristy Manning. 1997. *The Ecology of Place: Planning for Environment, Economy, and Community.* Washington, D.C.: Island Press.

Becker, Howard. 1995. "The Power of Inertia." *Qualitative Sociology* 18 (3): 301–310.

Berry, Brian J. L. 1977. *The Social Burdens of Environmental Pollution: A Comparative Metropolitan Data Source.* Cambridge, Mass.: Ballinger.

Berry, Wendell. 1995. "The Obligation of Care." *Sierra Magazine,* September/October, 62–67, 101.

Blumberg, Louis, and Robert Gottlieb. 1989. *War on Waste: Can America Win Its Battle with Garbage?* Washington, D.C.: Island Press.

Brecher, Jeremy, and Tim Costello. 1994. *Global Village or Global Pillage: Economic Reconstruction from the Bottom Up.* Boston: South End Press.

Brodkin, Evelyn Z. 1990. "Implementation as Policy Politics." In *Implementation and the Policy Process*, ed. Dennis J. Palumbo and Donald J. Calista, 107–118. New York: Greenwood.

Bryant, Bunyan, and Paul Mohai, eds. 1992. *Race and the Incidence of Environmental Hazards.* Boulder, Colo.: Westview.

Bullard, Robert D. 1990. *Dumping in Dixie: Race, Class, and Environmental Quality.* Boulder, Colo.: Westview.

Bullard, Robert D, ed. 1993. *Confronting Environmental Racism: Voices from the Grassroots.* Boston: South End Press.

_____. 1994. *Unequal Protection : Environmental Justice and Communities of Color.* San Francisco: Sierra Club Books.

Burchell, Robert W. 1997. "Economic and Fiscal Costs (and Benefits) of Sprawl." *The Urban Lawyer* 29 (2): 159–181.

Buttel, Frederick, and William L. Flinn. 1978. "Social Class and Mass Environmental Beliefs: A Reconsideration." *Environmental Behavior* 10:433–450.

Calavita, Nico. 1992. "Growth Machines and Ballot Box Planning: The San Diego Case." *Journal of Urban Affairs* 14:1–24.

California Coalition for Rural Housing. 1989. *Local Progress in Meeting the Low In-come Housing Challenge: A Survey of California Communities' Low Income Housing Production*. Sacramento: California Coalition for Rural Housing.

Capek, Stella M., and John I. Gilderbloom. 1992. *Community Versus Commodity: Tenants and the American City*. Albany: State University of New York Press.

Carnoy, Martin, and Derek Shearer. 1980. *Economic Democracy: The Challenge of the 1980s*. White Plains, N.Y.: M. E. Sharpe.

Castells, Manuel. 1985. "High Technology, Economic Restructuring, and the Urban-Regional Process in the United States." *Urban Affairs Annual Reviews* 28:11–40.

Cervero, Robert. 1988. "Paying for Off-Site Road Improvements Through Fees, Assessments, and Negotiations: Lessons from California." *Public Administration Review* 48 (1): 534–541.

Chiras, Daniel. 1995. *Voices for the Earth*. Boulder, Colo.: Johnson.

Christensen, Jon. 1997. "Cleaning Up Lake Tahoe Proves to Be Dirty Business." *High Country News*, May 12, 1.

Clark, Terry, and Edward Goetz. 1994. "The Antigrowth Machine: Can City Governments Control, Limit, or Manage Growth?" In *Urban Innovations: Creative Strategies for Turbulent Times*, ed. Terry Clark. Thousand Oaks, Calif.: Sage.

Clavel, Pierre. 1986. *The Progressive City: Planning and Participation, 1969–1984*. New Brunswick, N.J.: Rutgers University Press.

Cobb, Clifford, Ted Halstead, and Jonathan Rowe. 1995. *The Genuine Progress Indicator: Summary of Data and Methodology*. San Francisco: Redefining Progress.

Cohn, D'vera. 1979. "Big Is No Longer Beautiful for Many U.S. Communities." *Santa Barbara News Press*, March 4, A18.

Commission for Racial Justice. 1987. *Toxic Wastes and Race: A National Report on the Racial and Socioeconomic Characteristics of Communities with Hazardous Waste Sites*. New York: United Church of Christ.

Community Environmental Council. 1987. "Monitoring: The Missing Link in Effective Land-Use Decision-Making." Santa Barbara, Calif.: Community Environmental Council.

Condelli, Larry. 1980. "Challenge to the Growth Machine: The Santa Monica Rent Revolt." Unpublished paper, University of California, Santa Cruz.

Conners, Donald L., and Michael D. Bliss. 1988. *Exactions and Impact Fees: A Handbook for Real Estate Development*. Arlington, Va.: National Association of Industrial and Office Parks.

Construction Industry Federation. 1988. "CIF 1988 Regional Fee Survey." Brochure. San Diego, Calif.: Construction Industry Federation.

Construction Industry Research Board. 1971–1990. Building Permit Summary. Burbank, Calif.: Construction Industry Research Board. Photocopy.

Coyle, Kevin J. 1998. "Environmental Myths and Misconceptions." *The Polling Report* 14 (24): 2–4.

Cox, Kevin. 1992. "The Politics of Globalization: A Skeptic's View." *Political Geography* 11 (5): 427–429.

Creedon, Jeremiah. 1993. "The Power of Global Thinking?" *Utne Reader*, March/April, 22–26.

Cropper, Maureen L., and Wallace E. Oates. 1992. "Environmental Economics: A Survey." *Journal of Economic Literature* 30 (June): 675–740.

Cumberland, John. 1981. "Efficiency and Equity in Interregional Environmental Management." *Review of Regional Studies* 10 (2): 1–9.

Dahl, Robert Alan. 1961. *Who Governs?* New Haven: Yale University Press.

Daly, Herman E., and John B. Cobb. 1989. *For the Common Good: Redirecting the Economy Toward Community, the Environment, and a Sustainable Future.* Boston: Beacon.

Dalton, Keith. 1989. "Water Settlement OK'd." *Santa Barbara News Press,* June 9, A1.

Danielson, Michael, and James Doig. 1982. *New York: The Politics of Urban Regional Development.* Berkeley: University of California Press.

Deakin, Elizabeth. 1988. "Growth Controls and Growth Management: A Summary and Review of Empirical Research." Paper presented at UCLA extension conference on "The Growth Controversy in California: Searching for Common Ground," Manhattan Beach, Calif., June 16–17.

DeLeon, Richard E. 1992a. "The Urban Antiregime: Progressive Politics in San Francisco." *Urban Affairs Quarterly* 27 (4): 555–579.

———. 1992b. *Left Coast City: Progressive Politics in San Francisco, 1975–1991.* Lawrence: University Press of Kansas.

Dewey, John. 1954. *The Public and Its Problems.* Denver: Swallow.

Dimaggio, Paul, and Walter Powell. 1983. "The Iron Cage Revisited: Institutional Isomorphism and Collective Rationality in Organizational Fields." *American Sociological Review* 48 (1/6): 147–160.

Domhoff, G. William. 1983. *Who Rules America Now? A View for the 80s.* Englewood Cliffs, N.J.: Prentice-Hall.

Donovan, Todd, and Max Nieman. 1992. "Community Social Status, Suburban Growth, and Local Government Restrictions." *Urban Affairs Quarterly* 28 (2): 323–336.

Dowall, David E. 1980. "An Examination of Population-Growth-Managing Communities." *Policy Studies Journal* 9 (3): 414–427.

Dowall, David E., Marc Beyeler, and Chun-Cheung Sidney Wong. 1994. "Evaluation of California's Enterprise Zone and Employment and Economic Incentive Programs." Berkeley: California Policy Seminar, University of California.

Dowall, David E., and John Landis. 1982. *Land-Use Controls and Housing Costs: An Examination of San Francisco-Bay Area Communities.* Berkeley: Institute of Urban and Regional Development, University of California.

Duggan, Sharon, James G. Moose, and Tina Thomas. 1988. *Guide to the California Environmental Quality Act (CEQA).* 2d ed. Berkeley: Solano Press.

Duncan, Beverly, and Stanley Lieberson. 1970. *Metropolis and Region in Transition.* Beverly Hills: Sage.

Dunlap, Riley E. 1975. "The Socioeconomic Basis of the Environmental Movement: Old Data, New Data, and Implications for the Movement's Future." Paper presented at the annual meeting of the American Sociological Association, San Francisco, Calif., August.

Easton, Robert. 1972. *Black Tide: The Santa Barbara Oil Spill and Its Consequences.* New York: Delacorte Press.

Economic Research Associates. 1986. "Santa Barbara Economic Forecast and Hotel/Tourism Study." Prepared for City of Santa Barbara, California. San Francisco: Economic Research Associates.

———. 1990. "Fiscal Impacts of New Development: An Addendum to Report I, Existing Conditions and Issues." Prepared for City of Santa Monica, California, Department of Community and Economic Development.

Edelman, Murray. 1964. *The Symbolic Uses of Politics.* Urbana: University of Illinois Press.

Eisinger, Peter K. 1988. *The Rise of the Entrepreneurial State.* Madison: University of Wisconsin Press.

Elkin, Stephen. 1987. *City and Regime in the American Republic.* Chicago: University of Chicago Press.

Ellickson, Robert C. 1977. "Suburban Growth Controls: An Economic and Legal Analysis." *Yale Law Journal* 86: 385–511.

Elliot, M. 1981. "The Impact of Growth Control Regulations on Housing Prices in California." *Journal of the American Real Estate and Urban Economics Association* 5: 115–133.

Esteva, Gustavo, and Madhu Suri Prakash. 1998. *Grassroots Post-Modernism: Remaking the Soil of Cultures.* London: Zed Books.

Fasenfest, David. 1986. "Community Politics and Urban Redevelopment: Poletown, Detroit, and General Motors." *Urban Affairs Quarterly* 22 (1): 101–123.

Feagin, Joe R. 1983. *The Urban Real Estate Game.* Englewood Cliffs, N.J.: Prentice-Hall.

———. 1998. *The Free Enterprise City.* New Brunswick, N.J.: Rutgers University Press.

Feiock, Richard. 1991. "The Effects of Economic Development Policy on Local Growth." *American Journal of Political Science* 35 (3): 643–655.

———. 1994. "The Political Economy of Growth Management." *American Politics Quarterly* 22: 208–220.

Feldman, Thomas, and Andrew E. G. Jonas. 1999. "Sage Scrub Revolution? Property Rights, Political Fragmentation, and Conservation Planning in Southern California Under the Federal Endangered Species Act." Unpublished manuscript, University of Hull, England.

Fellmeth, Robert C. 1973. *The Politics of Land: Ralph Nader's Study Group Report on Land Use in California.* New York: Grossman Publishers.

Ferman, Barbara. 1996. *Challenging the Growth Machine: Neighborhood Politics in Chicago and Pittsburgh.* Lawrence: University Press of Kansas.

Feshbach, Murray, and Alfred Friendly Jr. 1992. *Ecocide in the U.S.S.R.: Health and Nature Under Siege* New York: Basic Books.

Fischel, William A. 1990. *Do Growth Controls Matter? A Review of Empirical Evidence on the Effectiveness and Efficiency of Local Government Land Use Regulation.* Cambridge, Mass.: Lincoln Institute of Land Policy.

Foglesong, Richard. 1990. "Magic Town: Orlando and Disney." Paper presented at the annual meeting of the Western Political Science Association, Newport Beach, Calif., March 22–24.

Frank, James E., and Paul B. Downing. 1988. "The Patterns of Impact Fee Use." In *Development Impact Fees: Policy Rationale, Practice, Theory, and Issues,* ed. Arthur C. Nelson, 3–21. Chicago: Planners' Press, American Planning Association.

Frech, H. E., and R. N. Lafferty. 1984. "The Effects of the California Coastal Commission on Housing Prices." *Journal of Urban Economics* 16: 105–123.

Freudenburg, William. 1993. "A 'Good Business Climate' as Bad Economic News." *Society and Natural Resources* 3 (4): 313–330.

Freudenburg, William, and Robert Gramling. 1994. "Bureaucratic Slippage and Failures of Agency Vigilance: The Case of the Environmental Studies Program." *Urban Affairs Quarterly* 41 (2): 214–239.

Frieden, Bernard. 1980. *The Environmental Protection Hustle.* Cambridge, Mass.: MIT Press.

Friedland, Roger. 1982. *Power and Crisis in the City.* London: Macmillan.

Friedland, Roger, Frances Piven, and Robert Alford. 1978. "Political Conflict, Urban Structure, and the Fiscal Crisis." In *Comparing Urban Policies,* ed. Douglas Ashford, 175–225. Beverly Hills, Calif.: Sage.

Fulton, William B. 1985. "Development Agreements in Santa Monica." Master's thesis in Architecture and Planning, University of California, Los Angeles.

Gebhard, David. 1990. "The Community as Client: Architectural Review in America." *Architecture California* 12 (1): 3–8.

Gebhard, D., and R. Winter. 1977. *A Guide to Architecture in Los Angeles and Southern California.* Santa Barbara, Calif.: Peregrine Smith.

German, Brad. 1990. "Winning at the Polls: New Rules for the 90s." *Builder*, June, 46–48.

Giddens, Anthony. 1985. *The Nation-State and Violence.* Berkeley and Los Angeles: University of California Press.

Glickfeld, Madelyn, LeRoy Graymer, and Kerry Morrison. 1987. "Trends in Local Growth Control Ballot Measures in California." *Journal of Environmental Law* 6 (87): 111–158.

Glickfeld, Madelyn, and Ned Levine. 1990. *The New Land Use Regulation "Revolution": Why California's Local Jurisdictions Enact Growth Control and Management Measures.* Los Angeles: University of California.

Gottdiener, Mark. 1983. "Some Theoretical Issues in Growth Control Analysis." *Urban Affairs Quarterly* June: 565–572.

———. 1987. *The Decline of Urban Politics.* Beverly Hills: Sage Publications.

Gottdiener, Mark, and Max Neiman. 1981 "Characteristics of Support for Local Growth Control." *Urban Affairs Quarterly* 17 (1): 55–73.

Gottdiener, Mark, and Joe R. Feagin. 1988. "The Paradigm Shift in Urban Sociology." *Urban Affairs Quarterly* 24 (2):163–187.

Grabher, G. 1990. "On the Weakness of Strong Ties: The Ambivalent Role of Inter-Firm Relations in the Decline and Reorganization of the Ruhr." Paper for Unit Labour Market and Employment (IIM), Wissenschaftszentrum (WZB), Berlin.

Graves, Gregory R., and Sally L. Simon, eds. 1980. *A History of Environmental Review in Santa Barbara County, California.* Santa Barbara: Graduate Program in Public Historical Studies, University of California.

Hamilton, Cynthia. 1990. "Women, Home, and Community: The Struggle in an Urban Environment." In *Reweaving the World: The Emergence of Ecofeminism*, ed. Irene Diamond and Gloria Orenstein. San Francisco: Sierra Club Books.

Hamilton, Rabinovitz, and Szanton, Inc. 1982. "Office Development in Santa Monica: The Municipal Fiscal and Housing Impact." Los Angeles: Hamilton, Rabinovitz, and Szanton, Inc. Photocopy.

Hawken, Paul. 1993. *The Ecology of Commerce: A Declaration of Sustainability*. New York: HarperBusiness.

Hayes, Dennis. 1995. "Transportation, Environment, and Innovation: An Interview with Dennis Hayes." *On the Ground* 1 (3): 1–3.

Heskin, Allan. 1983. *After the Battle Is Won: Political Contradictions in Santa Monica.* Los Angeles: Graduate School of Architecture and Planning, UCLA.

Hill, Christian. 1978. "Keeping the Lid On." *Wall Street Journal,* February 8, 1, 27.

Hill, Richard Child. 1974. "Separate and Unequal: Government Inequality in the Metropolis." *American Political Science Review* 68 (December): 1557–1568.

Hoch, Charles. 1984. "City Limits: Municipal Boundary Formation and Class Segregation." In *Marxism in the Metropolis,* ed. William K. Tabb and Larry Sawyers, 101–119. 2d ed. New York: Oxford University Press.

Hocking, Martin B. 1991. "Paper Versus Polystyrene: A Complex Choice." *Science,* February 1, 504–505.

Horner, Chris. 1990. "Welcome to La-La Land." *Pipeline and Utilities Construction,* November 4, 1.

Humphrey, Craig R., and Richard Krannich. 1980. "The Promotion of Growth in Small Urban Places and Its Impact on Population Change, 1975–1978." *Social Science Quarterly* 61 (3/4): 581–594.

Hundley, Norris, Jr. 1992. *The Great Thirst: Californians and Water, 1770s–1990s.* Berkeley and Los Angeles: University of California Press.

Hunter, Floyd. 1953. *Community Power Structure: A Study of Decision Makers.* Chapel Hill: University of North Carolina Press.

Huntington, Samuel P. 1952. "The Marasmus of the ICC: The Commission, the Railroads, and the Public Interest." *Yale Law Journal* 61: 467–509.

Janczyk, J. T., and W. C. Constance. 1980. "Impacts of Building Moratoria on Housing Markets Within a Region." *Growth and Change* 11: 11–19.

Jencks, Christopher. 1994. *The Homeless.* Cambridge, Mass.: Harvard University Press.

Jonas, Andrew. 1997. "Regulating Suburban Politics: 'Suburban-Defense Transition,' Institutional Capacities, and Territorial Reorganization in Southern California." In *Reconstructing Urban Regime Theory: Regulating Urban Politics in a Global Economy,* ed. Mickey Lauria. Thousand Oaks, Calif.: Sage.

Jonas, Andrew, and David Wilson, eds. forthcoming. *Twenty Years Later: Critical Perspectives on the Growth Machine Thesis.* Albany: State University of New York Press.

Kaiser, Edward K., Raymond J. Burby, and David H. Moreau. 1988. "Local Governments' Use of Water and Sewer Impact Fees and Related Policies: Current Practice in the Southeast." In *Development Impact Fees: Policy Rationale, Practice, Theory, and Issues,* ed. Arthur C. Nelson. Chicago: Planners Press.

Katz, L., and K. T. Rosen. 1980. "The Effects of Land-Use Controls on Housing Prices." Working paper no. 80.13, Center for Real Estate and Urban Economics, University of California, Berkeley.

_____. 1987. "The Interjurisdictional Effects of Growth Controls on Housing Prices." *Journal of Law and Economics* 30: 149–160.

Keating, W. Dennis. 1986. "Linking Downtown Development to Broader Community Goals: An Analysis of Linkage Policy in Three Cities." *Journal of the American Planning Association* 52 (2): 133–141.

Kemmis, Daniel. 1990. *Community and the Politics of Place.* Norman: University of Oklahoma Press.

Kmenta, Jan. 1986. *Elements of Econometrics.* 2d ed. New York: Macmillan.

Krannich, Richard S., and Craig R. Humphrey. 1983. "Local Mobilization and Community Growth: Toward an Assessment of the 'Growth Machine' Hypothesis." *Rural Sociology* 48 (1): 60–81.

Krauss, Celene. 1989. "Community Struggles and the Shaping of Democratic Consciousness." *Sociological Forum* 4 (2): 227–239.

Kristof, Kathy M. 1989. "Housing Affordability Rises Outside L.A., Orange County." *Los Angeles Times,* December 6, D1.

Kroeber, Alfred L. 1925. *Handbook of the Indians of California.* Smithsonian Institution Bureau of American Ethnology, Bulletin 78. Washington, D.C.: GPO. Reprint, New York: Dover Publications, 1976.

Landis, John. 1992. *Do Growth Controls Work? An Evaluation of Local Growth Control Programs in Seven California Cities.* Berkeley: California Policy Seminar, University of California.

Landis, John, Rolf Pendall, Robert Olshasky, and William Huang. 1995. "Fixing CEQA: Options and Opportunities for Reforming the California Environmental Quality Act." *CPS Brief* 9 (12).

Lash, Scott, and John Urry. 1994. *Economies of Signs and Space.* London: Sage.

Lategola, Amy Rose. 1992. "Who's Giving to Whom? An Analysis of Local Campaign Contributions and Competing Development Interests in Santa Barbara, California." Master's thesis, Department of Sociology, University of California, Santa Barbara.

Latour, Bruno. 1987. *Science in Action.* Cambridge: Harvard University Press.

Lauber, V. 1978. "Ecology and Elitism in American Society: The Fallacy of the Post-Materialist Hypothesis." Paper presented at the 32nd annual meeting of the Western Political Science Association, Los Angeles, California, March 16–18.

Lauria, Mickey, ed. 1997. *Reconstructing Urban Regime Theory: Regulating Urban Politics in a Global Economy.* Thousand Oaks, Calif.: Sage.

Lefebvre, Henri. 1991. *The Production of Space,* trans. Donald Nicholson-Smith. Oxford: Blackwell.

Leonard, H. Jeffrey. 1988. *Pollution and the Struggle for the World Product.* Cambridge, U.K.: Cambridge University Press.

Levy, Stephen. 1988. "California Economic Growth: Regional Market Update and Projections." Palo Alto, Calif.: Center for Continuing Study of the California Economy.

———. 1992. *Outlook for the California Economy.* Palo Alto, Calif.: Center for the Continuing Study of the California Economy.

———. 1994. *California Economic Growth.* Palo Alto, Calif.: Center for the Continuing Study of the California Economy.

Li, Mingche M., and H. James Brown. 1980. "Micro-Neighborhood Externalities and Hedonic Housing Prices." *Land Economics* 56 (May): 125–141.

Lindblom, Charles E. 1965. *The Intelligence of Democracy: Decision Making Through Mutual Adjustment.* New York: Free Press.

———. 1977. *Politics and Markets: The World's Political Economic Systems.* New York: Basic Books.

Logan, John R. 1978. "Growth, Politics, and the Stratification of Places." *American Journal of Sociology* 87 (2): 404–415.

Logan, John R., and Harvey L. Molotch. 1987. *Urban Fortunes: The Political Economy of Place.* Berkeley: University of California Press.

Logan, John, and Min Zhou. 1989. "Do Suburban Growth Controls Control Growth?" *American Sociological Review* 54 (3): 461–471.

_____. 1990. "The Adoption of Growth Controls in Suburban Communities." *Social Science Quarterly* 71 (1): 118–129.

Logan, John, and Todd Swanstrom, ed. 1990. *Beyond the City Limits.* Philadelphia: Temple University Press.

Logan, John R., Rachel Whaley, and Kyle Crowder. 1997. "The Character and Consequences of Growth Regimes: An Assessment of 20 years of Research." *Urban Affairs Review* 32 (5): 603–630.

Los Angeles Times. 1990. "Good Neighbor Policy." April 22, K1.

Lowi, Theodore J. 1969. *The End of Liberalism.* New Haven: Norton.

Lyon, Larry, Lawrence G. Felice, M. Ray Perryman, and E. Stephen Parker. 1981. "Community Power and Population Increase: An Empirical Test of the Growth Machine Model." *American Journal of Sociology* 81 (2): 235–260.

March, James G., and Johan P. Olsen. 1976. *Ambiguity and Choice in Organizations.* Bergen, Norway: Universitetsforlaget.

Mark, Jonathan H., and Michael A Goldberg. 1981. "Land Use Controls: The Case of Zoning in Vancouver." *Areuea Journal* 9 (winter): 418–436.

Markusen, Ann. 1978. "Class, Rent, and Sectoral Conflict: Uneven Development in Western U.S. Boomtowns." *Review of Radical Political Economics* 10 (3): 117–129.

Martin, Hugo. 1991. "8 Cities Repeatedly Overstep Growth Limits Set by Law." *Los Angeles Times,* August 25, B1.

Mazmanian, Daniel, and Paul Sabatier. 1983. *Implementation and Public Policy.* Glenview, Ill.: Scott, Foresman.

McConnell, Grant. 1966. *Private Power and American Democracy.* New York: Alfred A. Knopf.

McConnell, Virginia D., and Robert M. Schwab. 1990. "The Impact of Environmental Regulation in Industry Location Decisions: The Motor Vehicle Industry." *Land Economics* 66 (February): 67–81.

McEvoy, Arthur F. 1986. *The Fisherman's Problem: Ecology and Law in the California Fisheries, 1950–1980.* Cambridge, U.K.: Cambridge University Press.

Meadows, Donella H., Dennis L. Meadows, and Jorgen Randers. 1992. *Beyond the Limits: Confronting Global Collapse, Envisioning a Sustainable Future.* Mills, Vt.: Chelsea Green.

Mercer, Lloyd J., and W. Douglas Morgan. 1982. "An Estimate of Residential Growth Controls' Impact on House Prices." In *Resolving the Housing Crisis: Government Policy, Decontrol, and the Public Interest,* ed. M. Bruce Johnson, 189–215. San Francisco: Pacific Institute for Public Policy Research.

Meyer, Stephen M. 1993. "Environmentalism and Economic Prosperity: An Update." Unpublished manuscript, Cambridge, Mass., Department of Political Science, Massachusetts Institute of Technology.

_____. Forthcoming. *Environmentalism and Economic Prosperity.* Cambridge, Mass.: MIT Press.

Milbrath, Lester. 1965. *Political Participation: How and Why Do People Get Involved in Politics?* Skokie, Ill.: Rand McNally.

Mill, John Stuart. 1974. *On Liberty.* 1859. Reprint, Baltimore: Penguin.

Molotch, Harvey. 1976. "The City as a Growth Machine." *American Journal of Sociology* 82 (2): 309–330.

_____. 1990. "Urban Deals in Comparative Perspective." In *Beyond the City Limits*, ed. John Logan and Todd Swanstrom, 175–198. Philadelphia: Temple University Press.

_____. 1995. "Art in Economy: How Aesthetics and Design Build Los Angeles." *Competition and Change: The Journal of Global Business and Political Economy* 1 (2): 145–185.

_____. Forthcoming. "Growth Machine Links: Up, Down, and Across." In *Twenty Years Later: Critical Perspectives on the Growth Machine Thesis,* ed. Andrew Jonas and David Wilson. Albany: State University of New York Press.

Molotch, Harvey, and John R. Logan. 1984. "Tensions in the Growth Machine." *Social Problems* 31 (5): 483–499.

Molotch, Harvey, and Serena Vicari. 1989. "Three Ways to Build: The Development Process in the United States, Japan, and Italy." *Urban Affairs Quarterly* 24 (2): 188–214.

Molotch, Harvey, and Hugh Louch. 1994. "Santa Barbara County During the Recession: How Did Our Economy Do?" CORI Research Report 94 (1). Santa Barbara: Community and Organization Research Institute, University of California, Santa Barbara.

Molotch, Harvey, John Woolley, and Terry Jori. 1998. "Growing Firms in Declining Fields: Unanticipated Impacts of Oil Development." *Society and Natural Resources: An International Journal* 11 (2): 137–156.

Molotch, Harvey, William Freudenberg, and Krista Paulsen. 1999. "History Repeats Itself, But How?" Unpublished paper, Department of Sociology, University of California, Santa Barbara.

Muller, Thomas. 1975. *Growing and Declining Urban Areas: A Fiscal Comparison.* Washington, D.C.: Urban Institute.

Navarro, Peter, and Richard Carson. 1991. "Growth Controls: Policy Analysis for the Second Generation." *Policy Sciences* 24 (2): 127–152.

Neiman, Max. 1990. "Growth Control Project: The Mosaic of Intentions Propelling Regulation of Residential Development." Unpublished paper, University of California, Riverside.

Nelson, Arthur C. 1986 "Using Land Markets to Evaluate Urban Containment Programs." *APA Journal.* 52 (2): 156–171.

Noyelle, Thierry, and Thomas Stanback. 1984. *The Economic Transformation of American Cities.* Totowa, N.J.: Rowman and Allanheld.

Office of Planning and Research. 1986. *CEQA, California Environmental Quality Act: Statutes and Guidelines.* Sacramento, Calif.: Office of Planning and Research.

Parker, Morgan. 1991. "New Resort Will Receive Warm Welcome." *Santa Barbara News Press,* December 14, B3.

Patterson, Tom. 1971. *A Colony for California: Riverside's First Hundred Years.* Riverside, Calif.: Press-Enterprise Company.

Peterson, Paul. 1981 *City Limits.* Chicago: University of Chicago Press.

Piller, Charles. 1991. *The Fail-Safe Society: Community Defiance and the End of American Technological Optimism.* New York: Basic Books.

Pillsbury, R. 1990. *From Boarding House to Bistro.* Boston: Unwin Hyman.

Pincetl, Stephanie S. 1992. "The Politics of Growth Control: Struggles in Pasadena, California." *Urban Geography* 13 (5): 450–467.

_____. Forthcoming. "The Politics of Influence: Democracy and the Growth Machine in Orange County." In *Twenty Years Later: Critical Perspectives on the Growth Machine Thesis*, ed. Andrew Jonas and David Wilson. Albany: State University of New York Press.

Pindyck, Robert S., and Daniel L. Rubinfeld. 1981. *Econometric Models and Economic Forecasts*. New York: McGraw-Hill.

Platt, Rutherford H. 1996. *Land Use and Society: Geography, Law, and Public Policy*. Washington, D.C.: Island Press.

Plotkin, Sidney. 1987. *Keep Out: The Struggle for Land Use Control*. Berkeley: University of California Press.

Popper, Frank J. 1981. *The Politics of Land-Use Reform*. Madison: University of Wisconsin Press.

Porter, Douglas, and Lindell L. Marsh, eds. 1989. *Development Agreements: Practice, Policy, and Prospects*. Washington, D.C.: The Urban Land Institute.

Porter, Michael. 1990. *The Competitive Advantage of Nations*. The Free Press.

Power, Thomas Michael. 1992. *The Economic Pursuit of Quality*. Armonk, N.Y.: M. E. Sharpe.

Pressman, Jeffrey L., and Aaron Wildavsky. 1973. *Implementation: How Great Expectations in Washington Are Dashed in Oakland*. Berkeley: University of California Press.

Protash, William, and Mark Baldassare. 1983. "Growth Policies and Community Status: A Test and Modification of Logan's Theory." *Urban Affairs Quarterly* 18 (3): 397–412.

Putnam, Robert D. 1993. *Making Democracy Work: Civic Traditions in Modern Italy*. Princeton: Princeton University Press.

Rainey, James, and Mary Moore. 1996. "The Price of Success." *Los Angeles Times*, May 23, B1, 6.

Recht, Hausrath, and Associates. 1984. *The Economic Basis for an Office-Housing Production Program*. San Francisco: San Francisco Department of City Planning.

Rey, Sergio J., Jr. 1988. "Spatial Housing Markets and Growth Control Regulations: An Application of Alternative Econometric Approaches to Santa Barbara County, California." Master's thesis, University of California, Santa Barbara.

Riverside County. 1970–1989. *Yearly Analysis of Building Permits and Valuations*. Riverside, Calif.: Riverside County Department of Building and Safety. Photocopy.

Riverside County Planning Department. 1990. "Growth Management Element: Environmental Impact Report No. 350." Riverside, Calif.: Riverside County Planning Department.

Rose, Jerome G. 1979. *Legal Foundations of Land Use Planning*. New Brunswick, N.J.: Center for Urban Policy Research.

Rosenthal, Rob. 1994. *Homeless in Paradise: A Map of the Terrain*. Philadelphia: Temple University Press.

Rubin, Herbert J. 1989. "Symbolism and Economic Development Work: Perceptions of Urban Economic Development Practitioners." *American Review of Public Administration* 19 (3): 233–247.

Rudel, Thomas K. 1989. *Situations and Strategies in American Land-Use Planning.* Cambridge, U.K., and New York: Cambridge University Press.

Rudel, Thomas, and Samuel Richards. 1990. "Urbanization, Roads, and Rural Population Change in the Ecuadorian Andes." *Studies in Comparative International Development* 25 (3): 73–89.

Rudel Thomas K., and Bruce Horowitz. 1993. *Tropical Deforestation: Small Farmers and Land Clearing in the Ecuadorian Amazon.* New York: Columbia University Press.

Santa Barbara County. 1970–1989. *Calendar Year Reports.* Santa Barbara, Calif.: Santa Barbara County, Division of Building and Safety. Photocopy.

Santa Barbara News Press. 1990. "Developer to Build Around Landmark," January 26, B1.

_____. 1990. "Come and Join Mother Nature for Walk in New Preserve," February 10, B1.

Santa Monica Outlook. 1990. "Hotel Developer Seeks Ballot Help," July 5, A1.

Savage, Mike, and Alan Warde. 1993. *Urban Sociology, Capitalism, and Modernity.* New York: Continuum.

Scheiber, Harry N. 1973. "Urban Rivalry and Internal Improvements in the Old Northwest, 1820–1860." In *American Urban History,* ed. Alexander Callow Jr., 135–146. 2d ed. New York: Oxford University Press.

Schwartz, Seymour, David Hansen, Richard Green, William Moss and Richard Belzer. 1979. *The Effect of Growth Management on Housing Prices: Petaluma, California.* Davis, Calif.: Institute of Governmental Affairs.

Schwartz, Seymour I., Peter M. Zorn, and David E. Hansen 1986. "Research Design Issues and Pitfalls in Growth Control Studies." *Land Economics* 62 (3): 223–233.

Schwartz, Seymour I., and Peter M. Zorn. 1988. "A Critique of Quasi-Experimental and Statistical Controls for Measuring Program Effects: Application to Urban Growth Control." *Journal of Policy Analysis and Management* 7 (3): 491–505.

Scott, W. Richard. 1981. *Organizations: Rational, Natural, and Open Systems.* Englewood Cliffs, N.J.: Prentice-Hall.

_____. 1987. *Organizations: Rational, Natural, and Open Systems.* 2d ed. Englewood Cliffs, N.J.: Prentice-Hall.

Sedway Cook Associates. 1989. "Ocean Park Neighborhood Rezoning Alternatives." Santa Monica, Calif.: City of Santa Monica Planning Department.

Sharp, Kathleen. 1993 "A Battle Rages Down by the Sea." *Los Angeles Times,* May 5, A3, A19.

Shearer, Derek. 1982. "How the Progressives Won in Santa Monica." *Social Policy* 12 (3): 7–14.

Shiva, Vandana. 1988. *Staying Alive: Women, Ecology, and Development in India.* London: Zed Books.

Sites, William. 1997. "The Limits of Urban Regime Theory: New York City Under Koch, Dinkins, and Guiliani." *Urban Affairs Review* 32 (4): 536–557.

Skoro, Charles L. 1988. "Rankings of State Business Climates: An Evaluation of Their Usefulness in Forecasting." *Economic Development Quarterly* 2 (2): 138–152.

Smith, Michael Peter, and Marlene Keller. 1983. "Managed Growth and the Politics of Uneven Development in New Orleans." In *Restructuring the City*, ed. Susan Fainstein, 126–166. New York: Longman.

Snyder, Thomas P., and Michael A. Stegman. 1988. *Paying for Growth: Using Development Fees to Finance Infrastructure*. Chapel Hill: The Urban Land Institute, University of North Carolina at Chapel Hill.

Sollen, Robert. 1983 "Measure Defeat Followed Pattern of Other Elections." *Santa Barbara News Press*, December 10, D1–2.

_____. 1989. "The Santa Barbara Story." Unpublished manuscript, Santa Barbara, Calif.

Squires, George D., ed. 1989. *Unequal Partnerships: The Political Economy of Urban Redevelopment in Postwar America*. New Brunswick, N.J.: Rutgers University Press.

Stall, Bill. 1994. "Adviser Criticizes Brown's Portrayal of State Economy." *Los Angeles Times*, May 20, A3.

Stanton, Russ. 1989a. "Most Living Well in a County That Defies Stereotypes." *Riverside Press Enterprise*, July 26, AA24.

_____. 1989b. "One Million and Growing." *Press Enterprise*, July 26, AA1.

Starr, Chauncey. 1969 "Social Benefit Versus Technological Risk." *Science* 165 (September): 1232–1237.

Sternlieb, George, and James W. Hughes. 1983. *The Atlantic City Gamble*. Cambridge, Mass.: Harvard University Press.

Stone, Clarence. 1976. *Economic Growth and Neighborhood Discontent: System Bias in the Urban Renewal Program of Atlanta*. Chapel Hill: University of North Carolina Press.

_____. 1981. "Community Power Structure—A Further Look." *Urban Affairs Quarterly* 16 (4): 505–515.

_____. 1989. *Regime Politics: Governing Atlanta, 1946–1988*. Lawrence: University Press of Kansas.

Stroud, Nancy. 1988. "Legal Considerations of Development Impact Fees." In *Development Impact Fees*, ed. Arthur Nelson, 83–95. Chicago: Planners Press.

Summers, Gene F. 1976. "Small Towns Beware: Industry Can Be Costly." *Planning* 42 (May): 20–21.

Sunderland, T. A. 1989. "Auto Centers: Returning to a Birthplace." *Inland Business*, July, 12–20.

Swanstrom, Todd. 1985. *The Crisis of Growth Politics: Cleveland, Kucinich, and the Challenge of Urban Populism*. Philadelphia: Temple University Press.

Szasz, Andrew. 1994. *Ecopopulism: Toxic Waste and the Movement for Environmental Justice*. Minneapolis: University of Minnesota Press.

Tatum, Sandra. 1995. "Glitz and Growth Take a Major Hit in Santa Fe: Santa Fe Mayor, Debra Jaramillo: On the Job." *High Country News*, August 8.

Tiebout, Charles M. 1956. "A Pure Theory of Local Expenditures." *Journal of Political Economy* 64 (October): 416–424.

Tilly, Charles. 1981. *Big Structures, Large Processes, Huge Comparisons*. New York: Russell Sage Foundation.

Times Mirror. 1990. *The People, Press, and Politics: A Times Mirror Political Typology*. Washington, D.C.: Times Mirror Center for the People and the Press.

_____. 1994. *The People, The Press, and Politics : The New Political Landscape.* Los Angeles: Times Mirror Center for the People and the Press.

Tobey, James A. 1989. "The Impact of Domestic Environmental Policies on International Trade." Ph.D. dissertation, Department of Economics, University of Maryland, College Park.

Tocqueville, Alexis de. 1945. *Democracy in America.* 1835, 1840. Translated by Henry Reeve. New York: Vintage.

Tolbert, Pamela S., and Lynne G. Zucker. 1983. "Institutional Sources of Change in the Formal Structure of Organizations: The Diffusion of Civil Service Reform, 1880–1935." *Administrative Science Quarterly* 28 (1): 22–39.

Torres, Vicki. 1990. "Suit Challenges Developers' Role in Environmental Impact Reports." *Los Angeles Times,* July 10, B1.

Unger, James S. 1990. "Linkage Fees: Development Exactions for Social Investment." *Economic Development Commentary* 14 (1): 11–20.

Urban Land Institute and Gruen, Gruen, and Associates. 1977. *Effects of Regulation on Housing Costs.* Washington, D.C.: Urban Land Institute.

U.S. Census. 1970–1990. *Statistical Abstract of the United States.* Washington, D.C.: U.S. Department of Commerce.

U.S. Census. 1972, 1983, 1994. *County and City Data Book.* Washington, D.C.: U.S. Department of Commerce.

U.S. Department of Commerce. 1970–1989. *Construction Review: Bimonthly Industry Report.* January/February issues. Washington, D.C.: U.S. Department of Commerce, International Trade Administration.

Van Liere, Kent D., and Riley Dunlap. 1980. "The Social Bases of Environmental Concern: A Review of Hypotheses, Explanations, and Empirical Evidence." *Public Opinion Quarterly* 44 (2): 181–197.

Verba, Sidney, and Norman H. Nie. 1972. *Participation in America: Political Democracy and Social Equality.* New York: Harper and Row.

Vogel, Ronald K. 1992. *Urban Political Economy: Broward County, Florida.* Gainsville: University Press of Florida.

Vogel, Ronald K., and Bert E. Swanson. 1989. "The Growth Machine Versus the Antigrowth Coalition: The Battle for Our Communities." *Urban Affairs Quarterly* 25 (1): 63–85.

Walker, Richard, and Michael Heiman. 1981. "Quiet Revolution for Whom?" *Annals of the Association of American Geographers* 71 (1): 67–83.

Walton, John. 1966. "Substance and Artifact: The Current Status of Research on Community Power Structure." *American Journal of Sociology* 71 (1/6): 430–438.

_____. 1976. "Community Power and the Retreat for Politics: Full Circle After Twenty Years?" *Social Problems* 23 (3): 636–644.

_____. 1981. "The New Urban Sociology." *International Social Science Journal* 33: 374–390.

_____. 1992a. *Western Times and Water Wars: State, Culture, and Rebellion in California.* Berkeley: University of California Press.

_____. 1992b. "Urban Sociology: The Contribution and Limits of Political Economy." *Annual Review of Sociology,* Volume 19.

Warner, Kee. 1997. "Building Environmental Justice into Urban Growth." Paper presented at the 47th annual meeting of the Society for the Study of Social Problems. Toronto, Ontario, August 8–10.

Warner, Kee, and Harvey Molotch, with the collaboration of A. R. Lategola. 1992. "Growth Control: Inner Workings and External Effects." Berkeley: California Policy Seminar, University of California.

_____. 1995. "Power to Build: How Development Persists Despite Local Controls," *Urban Affairs Review,* 30(3): 378–406.

Warren, Roland. 1963. *The Community in America.* Chicago: Rand McNally.

Weiss, Kenneth R. 1996. "Ventura Considers Ban on Business Subsidies." *Los Angeles Times,* March 21, A3, A25.

Weston, I. P. 1990. "Parker to Seek Change of Venue on Suit." *Santa Barbara News Press,* April 27, B1.

White, George. 1988. "Southland Seen Leading 49 States in Economic Growth." *Los Angeles Times,* October 28, IV:1.

White, Kenneth J. 1988. "SHAZAM: A Comprehensive Computer Program for Regression Models (Version 6)." *Computational Statistics & Data Analysis,* December.

Whitt, J. Allen. 1982. *Urban Elites and Mass Transportation: The Dialectics of Power.* Princeton: Princeton University Press.

_____. 1988. "The Role of the Performing Arts in Urban Competition and Growth." In *Business Elites and Urban Development,* ed. Scott Cummings, 49–70. Albany: State University of New York Press.

Williams, Elizabeth A. 1995. "The Plan to Save Atlantic City: Consequences of Community Disarticulation on Central City Revitalization." Paper presented at the annual meeting of the American Sociological Association, Washington, D.C., August 19–23.

Wilson, Chris. 1997. *The Myth of Santa Fe: Creating a Modern Regional Tradition.* Albuquerque: University of New Mexico Press.

Wolch, Jennifer, and S. A. Gabriel. 1981. "Local Land Development Policies and Urban Housing Values." *Environment and Planning* 13: 1253–1276.

Wolkoff, Michael. 1983. "The Nature of Property Tax Abatement Awards." *Journal of the American Planning Association* 49 (Winter): 77–84.

Zorn, Peter M., Seymour I. Schwartz, and David E. Hansen. 1986. "Mitigating the Price Effects of Growth Control: Case Study of Davis, California." *Land Economics* 62 (1): 46–57.

Zukin, Sharon. 1980. "A Decade of the New Urban Research." *Theory and Society* 9: 575–601.

_____. 1991. *Landscapes of Power: From Detroit to Disney World.* Berkeley: University of California Press.

Index